JOURNAL OF
GREEN ENGINEERING

Volume 2, No. 4 (July 2012)

Special issue on

*Cognitive Radio and the Application to Green Communications:
Technical, Economic, and Regulatory Viewpoints*

Guest Editors:

Oliver Holland, Fernando J. Velez, Periklis Chatzimisios
and Arturas Medeisis

JOURNAL OF GREEN ENGINEERING

Chairperson: Ramjee Prasad, CTIF, Aalborg University, Denmark
Editor-in-Chief: Dina Simunic, University of Zagreb, Croatia

Editorial Board
Luis Kun, Homeland Security, National Defense University, i-College, USA
Dragan Boscovic, Motorola, USA
Panagiotis Demstichas, University of Piraeus, Greece
Afonso Ferreira, CNRS, France
Meir Goldman, Pi-Sheva Technology & Machines Ltd., Israel
Laurent Herault, CEA-LETI, MINATEC, France
Milan Dado, University of Zilina, Slovak Republic
Demetres Kouvatsos, University of Bradford, United Kingdom
Soulla Louca, University of Nicosia, Cyprus
Shingo Ohmori, CTIF-Japan, Japan
Doina Banciu, National Institute for Research and Development in Informatics, Romania
Hrvoje Domitrovic, University of Zagreb, Croatia
Reinhard Pfliegl, Austria Tech-Federal Agency for Technological Measures Ltd., Austria
Fernando Jose da Silva Velez, Universidade da Beira Interior, Portugal
Michel Israel, Medical University, Bulgaria
Sandro Rambaldi, Universita di Bologna, Italy
Debasis Bandyopadhyay, TCS, India

Aims and Scopes
Journal of Green Engineering will publish original, high quality, peer-reviewed research papers and review articles dealing with environmentally safe engineering including their systems. Paper submission is solicited on:

- Theoretical and numerical modeling of environmentally safe electrical engineering devices and systems.
- Simulation of performance of innovative energy supply systems including renewable energy systems, as well as energy harvesting systems.
- Modeling and optimization of human environmentally conscientiousness environment (especially related to electromagnetics and acoustics).
- Modeling and optimization of applications of engineering sciences and technology to medicine and biology.
- Advances in modeling including optimization, product modeling, fault detection and diagnostics, inverse models.
- Advances in software and systems interoperability, validation and calibration techniques. Simulation tools for sustainable environment (especially electromagnetic, and acoustic).
- Experiences on teaching environmentally safe engineering (including applications of engineering sciences and technology to medicine and biology).

All these topics may be addressed from a global scale to a microscopic scale, and for different phases during the life cycle.

JOURNAL OF GREEN ENGINEERING

Volume 2   No. 4   July 2012

Published, sold and distributed by:
River Publishers
P.O. Box 1657
Algade 42
9000 Aalborg
Denmark

Tel.: +45369953197
www.riverpublishers.com

Journal of Green Engineering is published four times a year.
Publication programme, 2011–2012: Volume 2 (4 issues)

ISSN 1904-4720

Editorial Foreword:
Cognitive Radio and the Application to Green Communications: Technical, Economic, and Regulatory Viewpoints*

Oliver Holland[1], Fernando J. Velez[2], Periklis Chatzimisios[3] and Arturas Medeisis[4]

[1] King's College London, U.K.; e-mail: oliver.holland@kcl.ac.uk
[2] Institute of Telecommunications, University of Beira Interior, Portugal; e-mail: fjv@ubi.pt
[3] Alexander Technological Educational Institution of Thessaloniki, Greece; e-mail: peris@it.teithe.gr
[4] Vilnius Gediminas Technical University, Lithuania; e-mail: arturas.medeisis@el.vgtu.lt

Demand for wireless data exchange is perpetually increasing, leading to pressure on deployed wireless capacity. As systems enhance their spectral efficiency to attempt to address this, evermore complex processing techniques are being employed. Such techniques increase power consumption, leading to greater heat output from processors in network elements and increasing powered cooling. Moreover, the requirement for wideband linearity of radio frequency circuitry in state-of-the-art OFDM systems, twinned with high transmission power capability on the network side and the inefficiencies of components designed to operate at such limits, leads to greater inefficiency hence heat power output and the need for even greater cooling. All of these ingredients significantly affect the power consumption of communications equipment. Indeed, merely considering raw transmission power,

*This special issue is organized by the COST Action IC0905 (COST-TERRA, www.cost-terra.org) and the ICT-ACROPOLIS Network of Excellence (www.ict-acropolis.eu).

the Shannon–Hartley theorem implies that necessary transmission power increases exponentially with required capacity.

Such power consumption issues can often be mitigated through cognitive radio (CR) techniques. The use of greater or more appropriate bandwidths through CR, twinned with the selection of appropriate modulation and in some cases the aggregation of bandwidths, as well as the opportunistic use of spectrum of more appropriate propagation characteristics or with better interference management, can reduce necessary transmission power for devices and systems. Moreover, CR techniques may achieve greater spatial/temporal awareness of connectivity options and their inherent power consumptions, as well as the power consumptions of chosen radio access characteristics. They might be able to use this awareness to dynamically switch to connections to better manage loads and interference, or might be able to minimize the power consumption of radio frequency circuitry through switching radio access technique given a required traffic load. At a simpler level, they may be able to appropriately select shorter range connectivity options minimizing path loss/shadowing hence necessary transmission power, in consideration of experienced/predicted traffic loads and channel characteristics. Such capabilities are particularly relevant considering that lower power equipment such as wireless LAN access points can often achieve far better efficiency than higher-power equipment such as macro-cell base stations. This manifestation is not only because of favorable path loss characteristics, but also because of the better efficiencies of the lower power components in such access points.

CR might also assist in making other areas of life more environmentally friendly. The application of CR to intelligent transportation, for example, facilitating the sharing of information in a timely manner and with appropriate data rates among vehicles and between vehicles and roadside infrastructure through enhancing the capabilities of systems such as IEEE 802.11p, could lead to the better management of traffic signals and better collaborative automated management of traffic speeds. This would limit unnecessary vehicular start/stop cycles and greatly reduce the burning of fossil fuels. Moreover, the presence of CR capability in a wide range of future machines will facilitate those machines talking to each other more ubiquitously than today. Such machine-to-machine communications could greatly enhance the energy efficiency of everyday living. One example of this is smart grids and smart homes, enabling the usage of energy locally, only when needed, and catering for the integration, better prediction of and capitalization from renewable resources. Among other benefits, such solutions would decrease wastage in the home through better management of appliances and heating/cooling options.

In order for the range of solutions described above to become reality, the deployment of CR technologies must be accelerated. A key challenge here is regulation: without regulatory rules permitting, many CR visions simply cannot exist. However, while regulations should be adapted to allow and facilitate CR, the chosen regulatory way forward must be both technically feasible from a primary user protection perspective and must provide for the commercial success of CR systems. Moreover, the immense social and economic benefit that might transpire through the emergence of CR must be maximized through the definition of appropriate regulations.

In view of the above observations, this special issue presents a number of papers covering aspects of the overlap between green communications and CR. In view that CR can significantly improve energy efficiency per se through offering more spectrum hence the reduction in interference through less frequency reuse, among other advantages, Tao Chen and co-authors investigate the optimal allocation of channels in a DSA scenario in view of minimizing the energy consumed per bit for secondary communications given a varying range of available channels. The presented optimization algorithm is shown to provide a tangible improvement in the throughput/energy ratio as compared with random allocation, in addition to various other benefits.

In a similar vein but delving more into the case of interference limited access, Enrico Del Re and co-authors discuss the allocation of resources in a cognitive radio scenario while minimizing energy consumption, and propose the introduction of two algorithms respectively based on water-filling and game theory. This is compared also with a simulated annealing approach. A significant improvement in energy efficiency is shown through the introduction of these algorithms.

It must be noted that despite yielding applications that can reduce energy consumption in communications, aspects of CR operation per se can imply additional energy consumption. In order to maximize the practicality of CR solutions it is important to minimize the associated energy consumption of CR operation. One key aspect of CR energy consumption is sensing. To this end, Reshma Syeda and Vinod Namboodiri discuss the energy efficiency of cooperative sensing in ad-hoc WLAN CR scenarios, whereby in the current regulatory realm WLANs are an interesting case for CR technology both in "conventional" license-exempt bands and in TV white space. This paper, as well as analyzing the energy efficiency of generic cooperative spectrum sensing schemes, proposes two new enhanced and adaptive schemes for these generic cases. Significant energy savings are shown through the generic cooperative sensing schemes, as well as additional savings for the enhanced and adaptive schemes.

In addition to CR being used as an energy saving solution in its own right, there are other important energy saving solutions that it can be twinned or might naturally combine with. One such solution is relaying and cooperative communications; significant transmission power can be saved through the deployment of relays as a coverage extension and capacity enhancement solution, and such relays might also be of simpler technology or powered by renewable resources, reducing their operational power consumption compared with additional base stations being deployed for frequency reuse. In line with such observations, Rasool Sadeghi and co-authors discuss the energy efficiency, capacity and delay implications of cooperative cognitive networks operating in a Wi-Fi (IEEE 802.11) context, enhancing the bounds of knowledge in this field and providing important recommendations.

Finally, given the benefits for energy efficiency along with other advantages through CR technologies, the importance of facilitating such technologies becomes clear. A key aspect of assessment of the potential for CR solutions is the understanding of characteristics in terms of interference profile and spectrum opportunities left by systems that are operating in the subject spectrum. To this end, Jan-Willem van Bloem and co-authors discuss the characteristics of Wi-Fi networks in terms of their spectrum utilization, congestion, and interference characteristics. This work is not only beneficial in terms of better understanding of such networks sharing spectrum together, it is also highly beneficial to the other types of systems aiming to share spectrum with Wi-Fi technologies.

As is evident from the above-mentioned contributions, many high-quality works have made it into this special issue. We are extremely thankful to the authors of the wide range of papers submitted to this issue, from which this select subset of contributions has been chosen. We hope you enjoy reading this special issue.

The Guest Editors

Energy Efficient Spectrum Access in Cognitive Wireless Access Networks

Tao Chen[1], Marja Matinmikko[1], Javier Del Ser[2] and Jianhua Zhang[3]

[1]*VTT Technical Research of Finland, FL-90571 Oulu, Finland;*
e-mail: {tao.chen,marja.matinmikko}@vtt.fi
[2]*TECNALIA Research & Innovation, E-20009 Donostia – San Sebastián, Spain;*
e-mail: javier.delser@tecnalia.com
[3]*Beijing University of Posts and Telecommunications, 100876 Beijing, China;*
e-mail: jhzhang@bupt.edu.cn

Received 10 May 2012; Accepted: 11 May 2012

Abstract

It is estimated that by 2020 the traffic in wireless access networks will increase drastically as compared to the 2010 level. To meet this challenge, it needs the huge increase of spectrum, very high spectrum efficiency, dense deployment of small cells, and highly energy efficient solutions. Combining these together, we believe that the cognitive wireless access network would be a promising solution. In this paper, we study the energy efficient spectrum access problem in dynamic spectrum access (DSA) based wireless access networks, in which densely deployed access points (AP) provide open access to mobile terminals (MT) by spectrum opportunities enabled by primary users (PU). The question is how the spectrum is allocated to MT via AP so that the bit/energy to deliver the data is maximized. We separate the problem into the channel selection problem of AP and the AP association problem of MT, and propose the distributed AP channel selection algorithm and MT association algorithm, which are run on the AP and MT separately, with a joint goal to improve the bit/energy delivery by the efficient use of spectrum. The proposed algorithms only rely on local information exchange to estimate inter-cell interference, and are therefore scalable to large networks. The per-

Journal of Green Engineering, Vol. 2, 305–328.

formance of the proposed algorithms are evaluated by simulation. It shows in average around 10% bit/energy improvement over the algorithms which randomly allocate channels and associate MTs.

Keywords: cognitive radio system, wireless access network, channel assignment, energy efficiency, interference control.

1 Introduction

Wireless communication services have become an integral part of people's everyday life. According to a collection of recent forecasts of the global mobile data traffic by the International Telecommunication Union Radiocommunication (ITU-R) in [4], there is an increasing data rate demand for the next decade (2012-2022). In fact, there has been a rapid uptake of smart phones, tablets and innovative mobile applications created by users which will lead to a significant increase in the mobile data traffic in [4], which was not predicted earlier in [1]. In light of the increasing data rates and growing numbers of new devices, the efficient use of the basic resources in wireless communications in terms of energy and spectrum become increasingly important design criteria.

In fact, a major challenge facing the mobile communication industry is to develop means to respond to the growing data rate demand while keeping the energy consumption at a reasonable level. The energy efficiency (EE) can be defined in different levels such as system level and component level as shown in [6]. Improvements in the energy efficiency of wireless networks can be obtained via interference control by dynamically switching to connections that can better optimize the resource usage. Cognitive radio systems (CRS) as defined by the International Telecommunication Union Radiocommunication sector (ITU-R), employ "technology that allows the system to obtain knowledge of its operational and geographical environment, established policies and its internal state; to dynamically and autonomously adjust its operational parameters and protocols according to its obtained knowledge in order to achieve predefined objectives; and to learn from the results obtained" [2]. The CRS techniques can be used in future wireless networks to improve the energy efficiency through interference control by assigning the frequency channels for operations by considering the interconnections of the throughput and the energy consumption.

We believe that the wireless access network based on dynamic spectrum access (DSA) in a densely deployed environment would be a promising solution to meet the future mobile traffic challenge with CRS capabilities. In such a network, not only the capacity improvement, but also EE are critical problems. The trade-off between capacity and EE in cellular networks has been shown in [8]. It is reported that in a multi-cell scenario the interference from neighboring cells will degrade both EE and spectrum efficiency (SE). Interference management becomes an important means to improve EE in DSA based wireless access networks. The challenges and solutions to achieve an energy efficient wireless communications are summarized in [11,13]. In [11], the cognitive radio (CR) approaches applied to the green cellular network are mentioned. However, less research has been done on the application of CR to energy efficient wireless networks.

In this paper we study the energy efficient resource allocation in a DSA based wireless access network. In such a network access points (AP) are densely deployed to provide open access to mobile terminals (MT). Spectrum of the network is opportunistically shared from primary users (PU). In multiple channels divided from available spectrum, an AP selects one working channel and all MTs attached to the AP use the same channel to communicate with the AP. In the previous work [7], the spectrum access problem in the targeted network has been studied, with the objective to improve the spectrum efficiency. We extend the work in [7] by taking into account EE in the channel allocation and MT association. The main contributions of the paper include: designing a local information exchange method to estimate inter-cell interference, proposing a local energy efficient metric used for distributed energy efficient algorithms, and developing the distributed energy efficient algorithms for AP channel selection and MT association, respectively.

The rest of this paper is organized as follows. The system model including the considered wireless access network and the EE metric is presented in Section 2. Section 3 describes the problem formulation for the selection of channels and access points (AP). The way to calculate the throughput of an AP based on given neighboring topology and traffic load of MTs is given in Section 4. Section 5 presents the proposed algorithm for the selection of channels for the APs. Section 6 presents the proposed algorithm for MTs to select the best AP. The performance of proposed algorithms is studied in Section 7. Some remarks regarding the proposed approach are given in Section 8. Finally, conclusions are drawn in Section 9.

2 System Model

We study wireless access networks under the DSA scheme, in which N APs provide Internet access to M MTs through spectrum chances opportunist-

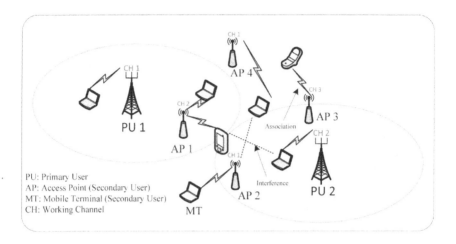

Figure 1 A DSA based wireless access network.

ically availed by P PU of spectrum. The studied network is illustrated in Figure 1. It can be applied to the scenarios where femtocells and Wi-Fi hot-spots are densely deployed with macrocells.

We use a simple spectrum coexistence model, which defines a reach range for each active PU. SUs out of the reach range of a PU are permitted to use the PU's licensed spectrum. The spectrum of the network is divided into C channels. For simplicity, we assume each channel has equal bandwidth W. Each SU i maintains an available channel list c_i, obtained from spectrum sensing, spectrum database, or other approaches defined for specific spectrum bands in question [14, 15]. We assume that an SU is able to update the available channel list timely according to activities of PUs.

An AP selects one channel out of c_i as its working channel. All MTs with the AP use the same working channel. The AP n and all its MTs form a cell n. In the following, depending on the context, the AP n may refer to the AP node or the cell n. An AP is allowed to shift its working channel according to its observation from the surrounding, e.g. if a PU is activated on its working channel, or if the AP finds another working channel that suffers from less interference from neighbors.

The time division multiple access (TDMA) is used by an AP. The channel access of an AP is structured into contiguous time frames, and each frame has L time slots. We assume APs are synchronized by time frames, which could be achieved by beacon or backhaul connections of APs. The detailed synchronization techniques are out of the scope of the paper.

An MT can associate with any AP, but only one AP at one time. Each MT m has its own downlink and uplink data rate preference, which are the up-limits of rates achieved by interference-free transmission. The MT can change its association if it finds another AP that can provide more energy efficient transmission to support its preferred data rate. Given the preferred rate R_m^D and R_m^U for the downlink and uplink of the MT m, the AP n assigns time slots according to the interference-free link rate, i.e. $r_m^D(n)$ to and $r_m^U(n)$ from the MT. The allocated slots are

$$l_m^D(n) = \left\lfloor \frac{R_m^D}{r_m^D(n)} \cdot L \right\rfloor \tag{1}$$

$$l_m^U(n) = \left\lfloor \frac{R_m^U}{r_m^U(n)} \cdot L \right\rfloor \tag{2}$$

where $\lfloor \cdot \rfloor$ is the ceil operator. We assume a node always has transmission in the allocated slots.

The following admission control rule is applied to the AP: if the unallocated time slots are not sufficient to support the preferred data rate of an MT, the AP will not admit the MT. As a result, not every MT can access to the network even it is in the reach range of an AP.

The interference-free link rate between the AP n and the MT m is computed by

$$r_m(n) = W \log_2 \left(1 + \frac{\gamma}{10^{\theta_{[dB]}/10}}\right) \tag{3}$$

where W is the bandwidth, γ is the signal to noise ratio, and θ means an operation θ dB from Shannon the capacity limit [10]. We set $\theta = 3$ dB. Additive white Gaussian noise (AWGN) channel is assumed to calculate $r_m(n)$.

The actual link rate is affected by interference from neighboring nodes on the same channel, and therefore makes the actual data rate less than the preferred data rate. We use the following way to estimate the actual data rate of an MT. For each AP and MT we define a reach range according to their transmission power. If a node is in the reach range of the other, the other node is its one-hop neighbor. A node fails to receive if any of its one-hop neighbors transmits simultaneously in the same channel. We assume the neighbors of a node which are two or more hops away will not affect the reception of the node. Consequently, we are able to use a collision model to take into account the interference. For each time slot, a collision probability is calculated based on the topology of local nodes and the scheduling on the time slot. The actual link rate as well as the downlink and uplink data rates of the MT are obtained thereafter.

To estimate the actual data rate, a node needs to discover its neighbors, and know the traffic load and scheduling of the neighbors. We assume that an effective neighbor discovery mechanism is used in the network, which allows MT and AP to know its local neighbors and their traffic load in the form of $l_m^D(n)$ and $l_m^U(n)$. By the local neighbor it means a neighbor two hops away. To simplify the analysis, we assume the AP randomly schedule the time slot to its MTs. That is, if the MT m is allocated $l_m^U(n)$ slots for uplink, the probability of the MT m to transmit at any slot is $p_m^U = l_m^U(n)/L$.

The activities of PUs as well as the mobility of SUs are assumed to be semi-stationary.

2.1 Power Consumption Model of AP and MT

We study the EE of the whole network, in which not only the energy consumed by radio parts, but also the energy used to support the network function of the node is taken into account. Since an AP is dedicated for the network access, the total energy of the AP is included in the study. For an MT we only consider the energy for the radio front-end as the significant portion of energy on MT is used to run applications.

The power consumption of an AP by building blocks is shown in Figure 2. As can be seen from this figure, the power consumption of the AP is insensitive to the transmission power. Hence, we define the power consumption model of an AP as follows:

$$P_A = P_0 + P_r \tag{4}$$

where P_A is the total power of an AP, P_0 is the power other than the radio power, and P_r is the radio power. Referring to Figure 2, we assume P_0 consumes $2/3$ of P_A, and P_r uses $1/3$ of P_A.

The power consumed by the radio front-end of an MT for transmission and reception are defined as P_T^M and P_R^M, respectively. The suggested values for P_T^M and P_R^M are referred to Chapter 20 of [12]. In this study, we set $P_T^M = 0.151$ W and $P_R^M = 0.148$ W.

The transmission power of an AP is set to P_t^A W for its MTs. The transmission power of an MT is set to P_t^M W.

In each downlink time slot, an AP transmits using the power of P_t^A, and consumes the total power of P_A. In the receiving and idle mode, the AP

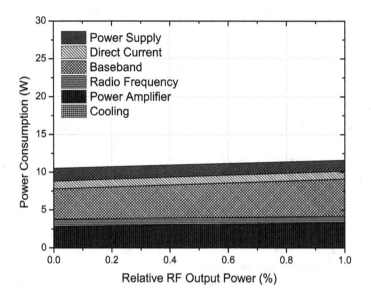

Figure 2 Power consumption of AP by building blocks as function of output power, redrawn from [5].

consumes the power of P_0. An MT consumes P_R^M in its downlink slots, P_T^M in its uplink slots, and 0 when it is idle. Note that we only consider the radio energy of an MT in this study.

2.2 EE Metric

EE of the whole network is measured by the ratio of the throughput of the whole network and the total energy consumption of the network, which is defined as:

$$EE = \frac{\sum_{n \in N} R(n)}{\sum_{n \in N} P(n)} \text{ bit/Joule} \qquad (5)$$

where EE is the network EE metric, $R(n)$ is the throughput of an AP n, and $P(n)$ is the average power consumed by the cell n. $R(n)$ is the sum of downlink and uplink throughput, and $P(n)$ includes the power of AP and its MTs.

In our proposed algorithm, an MT uses an EE metric to associate with an AP which offers the best bit/energy gain. As an MT can only obtain local information of its neighbors, it is not feasible to use EE from (5) in the algorithm. Instead, we use the local EE metric, which is defined as

$$ee(S_A) = \frac{\sum_{i \in S_A} R(i)}{\sum_{i \in S_A} P(i)} \text{ bit/Joule} \tag{6}$$

where $ee(S_A)$ is the local EE metric which takes into account the bit/energy cost of the cell set S_A. According to different energy efficient MT association algorithms, S_A may include the cell n of the MT m, and neighboring cells of the cell n, denoted as $N_A(n)$, or some neighboring cells or $N_A(n)$. We define $N_A(i)$ is the local neighboring cells of the node i. For the MT m on the AP n, we have $N_A(m) = N_A(n)$.

3 Problem Formulation

Let $R^D(n)$ and $R^U(n)$ be the actual downlink and uplink data rates of the AP n, $R(n)$ be the actual data rate of the AP n, $R_m^D(n)$ and $R_m^U(n)$ be the actual downlink and uplink data rate of the MT m on the AP n, $N_{PU}(i)$ be the neighboring PUs of the node i, $N(i)$ be the one-hop neighbors of the node i, and $c(n)$ be the available channels of the cell n. The problem of energy efficient spectrum access of the network is formulated as follows:

$$\max_N \quad EE = \frac{\sum_{n \in N} R(n)}{\sum_{n \in N} P(n)} \tag{7}$$

$$\text{subject to} \quad c(n) = \bigcap_{m \in n} c_m \cap c_n \text{ for } \forall n \in N \tag{8}$$

$$l_m^D(n) = \lfloor \frac{R_m^D}{r_m^D(n)} \cdot L \rfloor \text{ for } m \in \text{AP } n \tag{9}$$

$$l_m^U(n) = \lfloor \frac{R_m^U}{r_m^U(n)} \cdot L \rfloor \text{ for } m \in \text{AP } n \tag{10}$$

$$\sum_{m \in n} (l_m^D(n) + l_m^U(n)) \le L \text{ for } \forall n \in N \tag{11}$$

$$R(n) = R^D(n) + R^U(n) \text{ for } \forall n \in N \tag{12}$$

$$R^D(n) = f(R_m^D, R_m^U | m \in A(i | i \in N_A(n))) \tag{13}$$

$$R^U(n) = g(R_m^U, R_m^U | m \in A(i | i \in N_A(n))) \tag{14}$$

where $A(n)$ is the MTs of the AP n, $f(R_m^D, R_m^U | m \in A(i | i \in N_A(n)))$ and $g(R_m^U, R_m^U | m \in A(i | i \in N_A(n)))$ are the functions to compute the actual downlink and uplink data rate of the AP n in term of the traffic load and the topology of neighboring MTs and APs surrounding the AP n.

In the problem (7), (8) is the channel constraint of SUs, (9), (10) are the data rate constraints of an MT, (11) is the admission control of an AP, and (12), (13), (14) are the actual date rates of an AP affected by interference.

It is difficult to solve (7). We separate it into two sub-problems and use distributed iterative algorithms to approach the optimized result. One is the AP channel selection problem and the other is the MT association problem. Two problems are dealt with independently, and only local neighboring information is required to solve them.

In the AP channel selection problem, each AP estimates on different available channels the actual data rate that can be achieved under the current MT association and traffic load. The best channel provides maximum network throughput.

Since there is a network throughput gain to swift to the best channel and the power consumption of the network is not changed, the EE of the network is maximized. Instead of calculating $ee(S_A)$ metric of an AP, we use the throughput to get the most energy efficient channel. In this way the complexity of the problem is significantly reduced.

The problem is modeled as follows:

$$\underset{c \in c(n)}{\text{argmax}} \sum_{i \in N_A(n)} (R(i)|_c = R^D(i)|_c + R^U(i)|_c) \text{ for each AP } n \quad (15)$$

$$\text{subject to} \quad c(n) = \underset{m \in n}{\cap} c_m \cap c_n \text{ for } \forall n \in N \quad (16)$$

$$\sum_{m \in n} (l_m^D(n) + l_m^U(n)) \le L \text{ for } \forall n \in N \quad (17)$$

$$R^D(n) = f(R_m^D, R_m^U | m \in A(i | i \in N_A(n))) \quad (18)$$

$$R^U(n) = g(R_m^U, R_m^U | m \in A(i | i \in N_A(n))) \quad (19)$$

where $x|_c$ denotes the value of x on the channel c. Note that $\cdot|_c$ is omitted in (17), (18), (19). In the problem (15), (16) is the channel constraint of the AP, (17) is the admission control of the AP, and (18), (19) are the constraints to achieve actual data rates.

In the MT association problem, each MT evaluates the change of bit/energy of the local area when it attaches to different neighboring APs. The local area includes the cell of the MT and its neighboring cells. The AP with best bit/energy becomes the choice of the MT and the EE of the network is improved. The problem is modeled as the follows:

$$\underset{n \in N_A(m)}{\text{argmax}} \sum_{n \in S_A} ee(S_A) \text{ for each MT } m \qquad (20)$$

$$\text{subject to} \quad S_A = N_A(m) \cap a(m) \qquad (21)$$

$$\sum_{i \in n} (l_i^D(n) + l_i^U(n)) \leq L \text{ for } \forall n \in N \qquad (22)$$

$$R(n) = R^D(n) + R^U(n) \text{ for } \forall n \in N \qquad (23)$$

$$R^D(n) = f(R_i^D, R_i^U | i \in A(j | j \in N_A(n))) \qquad (24)$$

$$R^U(n) = g(R_i^U, R_i^U | i \in A(j | j \in N_A(n))) \qquad (25)$$

In the problem (20), (22) is the admission control of the AP, and (23), (24), (25) are the constraints to achieve actual date rates.

4 Throughput of AP

Given the interference-free link rate of $r_m^D(n)$ and $r_m^U(n)$, the allocated time slots of $l_m^D(n)$ and $l_m^U(n)$, the topology of neighboring nodes, and the random scheduling of time slots in each AP, we are able to calculate the collision probability on transmission at each time slot. With the collision probability, the throughput of an AP is obtained.

As shown in Figure 3, the collisions between two APs are divided into four basic cases:

- Case I: On the downlink of an MT, MTs of another AP collide the MT from their uplinks but the neighboring AP can not reach the MT;
- Case II: On the downlink of an MT, MTs of another AP collide the MT from their uplinks and the neighboring AP collides the MT from the downlink;
- Case III: On the uplink of an AP, MTs of another AP collide the AP from their uplinks but the neighboring AP can not reach the AP;
- Case IV: On the uplink of an AP, MTs of another AP collide the AP from their uplinks and the neighboring AP collides the AP from the downlink.

We first analyze the throughput of an AP, denoted as the AP a, on the channel c. The AP a has K MTs, and its neighboring AP b on the same channel has M MTs. In Case I of Figure 3, assuming an MT m of the AP a detects k one-hop neighboring MTs from the AP b, the probability that the AP b will

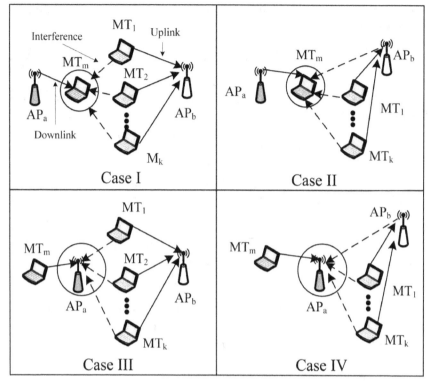

Figure 3 Interference in different topologies.

collide with the downlink transmission of the MT m on a time slot is

$$p_m^c(b) = \frac{l_m^D(a)}{L} \cdot \frac{\sum_{i \in N_k} l_i^U(b)}{L} \qquad (26)$$

where N_k is k one-hop neighboring MTs from the AP b.

In Case II, the downlink of the MT m contends with the downlink of the AP b and the uplink of its k MTs. The collision probability of the MT m with the downlink of the AP b is

$$\frac{l_m^D(a)}{L} \cdot \frac{\sum_{i \in A(b)} l_i^D(b)}{L}$$

and with k MTs is

$$\frac{l_m^D(a)}{L} \cdot \frac{\sum_{i \in N_k} l_i^U(b)}{L}$$

with the uplink of m MTs in AP b's cell. The collision probability of the MT i in this case is

$$p_m^c(b) = \frac{l_m^D(a)}{L} \cdot \frac{\sum_{i \in A(b)} l_i^D(b) + \sum_{i \in N_k} l_i^U(b)}{L} \qquad (27)$$

In the same way we can get the collision probabilities of the AP a in Cases III and IV as

$$p_a^c(b) = \frac{\sum_{i \in A(a)} l_i^U(a)}{L} \cdot \frac{\sum_{i \in N_k} l_i^U(b)}{L} \qquad (28)$$

$$p_a^c(b) = \frac{\sum_{i \in A(a)} l_i^U(a)}{L} \cdot \frac{\sum_{i \in A(b)} l_i^D(b) + \sum_{i \in N_k} l_i^U(b)}{L} \qquad (29)$$

respectively.

The collision probability of the node i with all neighboring cells on the channel c is

$$p_i^c = \bigcup_{b \in N_A(i)^c} p_i^c(b) \qquad (30)$$

where $N_A(i)^c$ is the neighboring AP list of the node i on the channel c, and the node i can be AP or MT.

The throughput of the AP a on the channel c is then

$$R^c(a) = \sum_{m \in A(a)} \frac{r_m^D(a) \cdot l_m^D(a) \cdot p_m^c + r_m^U(a) \cdot l_m^U(a) \cdot p_a^c}{L} \qquad (31)$$

where the MTs of the AP a are on channel c.

5 Energy Efficient Channel Selection of AP

In this section we describe the distributed channel selection algorithm. In the algorithm the AP n calculates its throughput on all available channels and gets a channel list $C'(n)$ on which its throughput is larger than that on the current working channel. For each channel in $C'(n)$, it calculates the throughput changes on all affected neighboring cells after the channel shift. The AP gets the throughput of its neighbor APs by information exchange through wireless links connected by MTs or wired links.

Note that we only calculate the throughput gain received by the AP n plus the throughput lost of all neighboring cells on the new channel, as the throughput gains of the neighboring APs on the current working channel of the AP have no impact on the channel selection.

From the channels in $C'(n)$, the AP n selects the one that can achieve the maximum throughput to shift, or in other word, it shifts to the channel:

$$c' = \underset{c' \in C'(n)}{\operatorname{argmax}} (\Delta R^{c'}(n) + \sum_{b \in N_A^{c'}(n)} \Delta \widetilde{R}^{c'}(b)) \tag{32}$$

where $\Delta R^{c'}(n)$ is the throughput gain of the AP n moving to the new channel c', and $\Delta \widetilde{R}^{c'}(b)$ is the throughput lost by the neighboring AP $b \in N_A^{c'}(n)$. $N_A^{c'}(n)$ are the neighboring APs on the channel c'.

To avoid the ping-pong effect, each time an AP n starts the algorithm with a probability of

$$p_A(n) = 1 - \beta \frac{R(n)}{\sum_{m \in A(n)} \frac{r_m^U(n) \cdot l_m^U(a) + r_m^D(n) \cdot l_m^D(a)}{L}} \tag{33}$$

over its working channel, where $\beta \in [0, 1]$ is a weight to control the sensitivity of the algorithm, and $p_A(n)$ is a simulated annealing factor to reduce the exploration when the current throughput is close to the interference-free throughput.

To reduce the complexity to calculate (32), the simplified version of the algorithm chooses the new channel based on

$$c = \underset{c' \in C'(n)}{\operatorname{argmax}} \Delta R^{c'}(a) \tag{34}$$

That is, an AP selects a new working channel without considering the throughput impact to the neighboring cells.

6 Energy Efficient MT Association

In this section we present the distributed MT association algorithm. An MT associates with an AP that can offer better bit/energy gain. The bit/energy is affected by the link rate between the MT and the AP as well as the topology of the neighboring nodes of the cell.

The energy efficient MT association algorithm is proposed as follows. An MT joins the first discovered AP. Once detecting a new AP, the MT will get by information exchange the MTs and the throughput of the AP. Moreover, we assume an MT will be informed by its AP when the neighboring cells have changes on the working channel or associated MTs. The MT will then gradually learn adjacent APs on its available channels. Assume that an MT m

initially joins the AP n, and later knows all neighboring APs and stores them in the list $N_A(m)$. Knowing the throughput of each AP in $N_A(m)$, the MT m calculates the bit/energy metric when accessing an neighboring AP. Since a re-association will only affect the bit/energy gain of neighboring cells, the affected cells when the MT m shifts from the AP n to the AP $b \in N_A(m)$ is $S_A(n, b) = n \cup b \cup N_A^{c(n)}(n) \cup N_A^{c(b)}(b)$. The MT m can get an AP list $S(m)$ satisfying

$$\delta_{ee}(n, b) = ee(S_A(n, b)) - ee'(S_A(n, b)) > 0 \qquad (35)$$

for $b \in S(m)$, where $\delta_{ee}(n, b)$ is the bit/energy gain for the re-association from the AP n to b, and $ee(S_A(n, b))$ and $ee'(S_A(n, b)$ are the bit/energy metric of S_A before and after the re-association of the MT m. Equation (35) ensures that the re-association does reduce the EE of the whole network.

The MT m selects the AP n where

$$n = \underset{b \in S(m)}{\operatorname{argmax}} \ \delta_{ee}(n, b) \qquad (36)$$

to re-associate with the probability:

$$p_M(m) = 1 - \alpha \frac{r_m^D(n) + r_m^U(n)}{\underset{b \in S(m)}{\max} (r_m^D(b) + r_m^U(b))} \qquad (37)$$

where $\alpha \in [0, 1]$.

Like in Section 5, the probability $p_M(m)$ acts as a simulated annealing factor to reduce the re-association frequency when the current link rate of the MT m is close to the highest.

To reduce the complexity of the algorithm, two simplified versions are proposed. The first one only has the AP n and b in S_A of (35). The second one is based on the first one but is even simple: it only uses the link rate to calculate the bit/energy metric $ee(n, b)$. The results from simplified version are suboptimal. However, the complexity of the algorithm is significantly reduced.

7 Simulation Study

The simulation is used to evaluate the performance of the proposed algorithms. The simulation setup is the following. A given number of PUs, APs and MTs are randomly placed on a 600 m×600 m playground. The maximum reach range of PUs are set to 200 m, and APs as well MTs are set to 100 m.

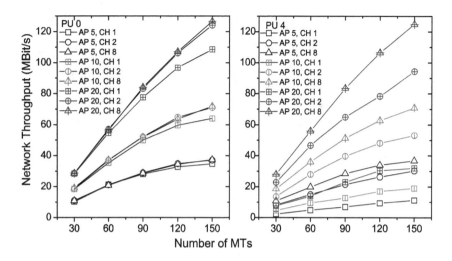

Figure 4 Network throughput under proposed algorithms.

The number of channels in the network are set to 1, 2, 4 or 8. Each channel has a bandwidth of 1 MHz. A PU randomly picks up one channel as its working channel. APs and MTs in the reach range of a PU avoid using its channel.

The power of an AP is set to 10 W, in which the radio power is 3.5 W and other parts consumes 6.5 W. The transmission power of an AP and an MT is set to 32 dBm. No power control is applied. The reciprocity of the downlink and uplink channel is assumed. Hence, the downlik and uplink interference-free link rates are identical.

Each MT has the downlink and uplink data rate preference of 500 kbit/s, respectively. The interference-free link rate between AP and MT is calculated by Equation (3) and the requested time slots are obtained thereafter. The path loss model from [9] is used:

$$l(\text{dB}) = 37 + 32 \log_{10}(r) \tag{38}$$

where r is the distance in meter. The noise in γ is calculated by

$$N = k \cdot T \cdot \text{NF} \cdot W \tag{39}$$

where $k = 1.3804 \cdot 10^{-23} J/K$ is the Boltzmann constant, $T = 290K$ is the temperature, NF $= 7$ dB is the noise figure, and W is the channel bandwidth.

The parameter α and β in Equations (37) and (33) are set to 0.8.

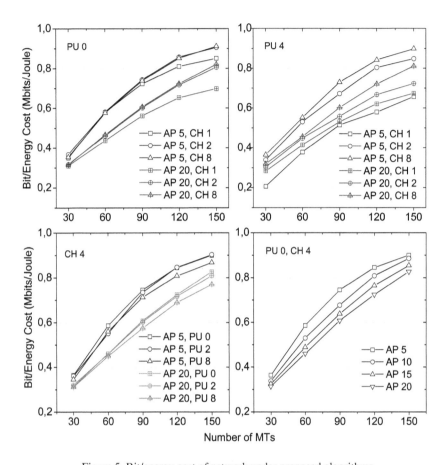

Figure 5 Bit/energy cost of network under proposed algorithms.

For each setup the simulation runs multiple times and the results are averaged.

Figure 4 shows the network throughput under different network setups. It can be seen from the figure, that the network throughput increases as more APs are deployed, because more APs increase the capacity of the network, and also get more MTs served. Figure 4 also shows the influence of available channels to the network throughput. More channels provide better throughput as the channel selection algorithm mitigates interference between neighbor cells. However, when there is no PU in the network, the throughput gain is not significant, especially when the available channels are more than one.

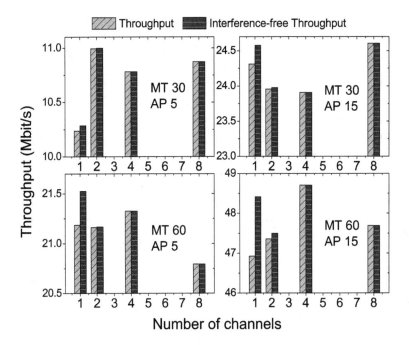

Figure 6 Network throughput and interference-free throughput under different available channels, no PU in network.

When there are 4 PUs in the network, more channels provide APs significant throughput improvements.

The bit/energy cost of the network is shown in Figure 5. Similar to the results in Figure 4, more MTs and more channels provide better bit/energy results. It is because the network is able to accommodate more throughput. We can see from this figure, especially from the right-bottom figure, that less APs provide higher EE. The reason is that each AP has significant power overhead compared to the MT. For the same available channels, the impact of the number of PUs is shown in the left-bottom graph of Figure 4. The number of PUs has less impact on EE of the network when there are less MTs, but the impact increases as MTs increase. It also shows 2 PUs and 8 PUs have similar effect on network EE.

It is interesting to see how the proposed algorithm exploits channel opportunities to avoid interference and then improves the throughput and EE. We use the interference-free throughput, which assumes that each AP has its own

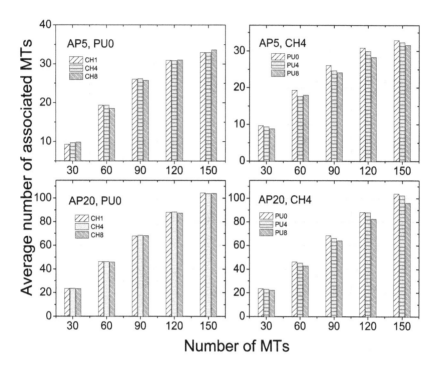

Figure 7 Average number of associated MTs in network.

channel and therefore no interference from other cells, as the reference for comparison. Shown in Figure 6, in the cases of 4 and 8 channels, those two throughput are the same, meaning that APs select channels in a way to avoid interference from other cells. When there is only one channel in the network, inter-cell interference is inevitable and it reduces the throughput. It is also shown in the figure that more MTs and more APs generate more interference.

Due to the reach range of an AP, as well as the capacity limit of an AP, not every MT can access an AP. Figure 7 shows the average number of associated MTs in the network. In the case of 5 APs, less than 30% of MTs gain access to APs. When there are 20 APs, around 70% of MTs gain access to APs. The reason is straightforward as more APs provide more coverage. But we can see that even if the number of APs is large, there is a significant number of MTs that can not access the network. The number of PUs affects the number of associated MTs. The more PUs are in the network, the less MTs can access the network. However, the impact of PUs on associated MTs is not severe.

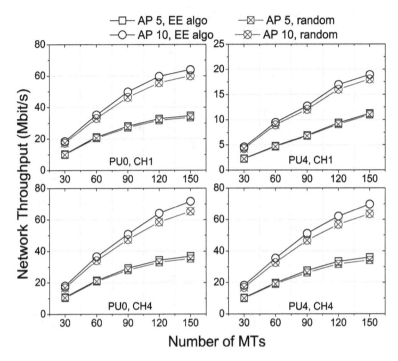

Figure 8 Comparison of network throughput between proposed algorithm and random association algorithm.

Finally, Figures 8 and 9 compare the throughput and EE of the network between proposed algorithm and a reference algorithm, which randomly associates MTs with nearby APs. The proposed algorithm improves the throughput and EE of the network as compared to the random algorithm. The gain of throughput over the random algorithm increases as the number of MTs, and/or the number of APs increase. For EE of the network, the gain of EE increases slower than throughput when the number of MTs increases. In average, 10% of EE gain can be achieved by proposed algorithm compared to the random algorithm.

8 Discussion

In this paper, MT association and AP channel selection algorithms are performed separately, but they affect each other. The re-association of an MT may change the load of the new AP, and in turn the interference to neigh-

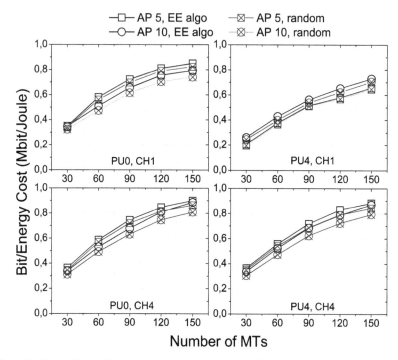

Figure 9 Comparison of bit/energy cost between proposed algorithm and random association algorithm.

boring cells. The channel change of an AP will change the neighbors of its MTs, and in turn let MTs to initialize re-association processes. The interaction between two algorithms move the network topology towards an optimal configuration.

We do not show the impact of dynamic behaviors of PUs on the performance of the algorithms. However, as the algorithms run in a distributed way, the changes of channel will be adapted by the network and in a long run the network will return to an optimal setup.

Strong assumptions on channel access and scheduling are applied in this paper. The reason is to simplify the analysis and provide insights on the problem. In practice, the collision probability between cells may be obtained from measurement. In this case the algorithms can be applied in practical wireless access systems.

Note that the inter-AP coordination will further improve throughput and EE of the network. Proper coordination among neighboring APs may significantly reduce the inter-cell interference. For instance, IEEE 802.16h uses joint scheduling among BSs to improve the spectrum reuse in the overlapping area of cells [3]. Future work could study efficient inter-AP communications for interference coordination and energy saving.

9 Conclusion

We study in a DSA based wireless access network the energy efficient spectrum access problem with the goal to maximize the bit/energy gain. The original problem is divided into two sub-problems, i.e. the channel selection problem of AP and the AP association problem of MT. We proposed distributed approaches to solve those two problems, which only rely on the local information exchange of inter-cell interference. The distributed AP channel selection algorithm and MT association algorithm are developed, which allow an AP to explore spectrum with less interference, and an MT to connect to an AP with the most bit/energy gain. Two algorithms are interactive in a way that allows an MT to access spectrum resource with less interference and more energy efficiently. Our approach is flexible and provides scalability to large networks. It can be applied to small cell environment where femtocells and Wi-Fi APs are densely deployed with macrocells.

Acknowledgements

This work has been partially supported by Finnish Funding Agency for Technology and Innovation, under the JADE project (DN 40474/09), Academy of Finland, under the SMAS project (DN 134624), and Spanish Ministry of Science and Innovation, under project number TEC2011-28250-C02-02.

References

[1] ITU-R Report M.2072, World Mobile Telecommunication Market Forecast, ITU-R, 2005.
[2] ITU-R Report SM.2152, Definitions of Software Defined Radio (SDR) and Cognitive Radio Ssystem (CRS), ITU-R, 2009.
[3] IEEE 802.16h-2010, Standard for Local and Metropolitan Area Networks, Part 16: Air Interface for Broadband Wireless Access Systems, Amendment 2: Improved Coexistence Mechanisms for License-exempt Operation, IEEE, 2010.
[4] ITU-R Report M.2243, Assessment of the Global Mobile Broadband Deployments and Forecasts for International Mobile Telecommunications, ITU-R, 2011.

[5] G. Auer et al. D2.3 Energy Efficiency Analysis of the Reference Systems, Areas of Improvements and Target Breakdown, 2010. Available at https://www.ict-earth.eu/publications/deliverables/deliverables.html, last accessed April 2012.

[6] E. Calvanese Strinati, A. De Domenico, and L. Herault. Green communications: An emerging challenge for mobile broadband communication networks. *Journal of Green Engineering*, 1(3):267–301, 2011.

[7] T. Chen, H. Zhang, M. Hoyhtya, and M. D. Katz. Spectrum self-coexistence in cognitive wireless access networks. In *Proceedings IEEE Globecom 2009*, pages 1–6, 2009.

[8] Y. Chen, S. Zhang, S. Xu, and G.Y. Li. Fundamental trade-offs on green wireless networks. *IEEE Communications Magazine*, 49(6):30–37, June 2011.

[9] D.J. Cichon and D.J. Kurner. EURO-COST 231 Final Report, 1998.

[10] H. Claussen. Performance of macro- and co-channel femtocells in a hierarchical cell structure. In *Proceedings of IEEE 18th International Symposium on Personal, Indoor and Mobile Radio Communications (PIMRC)*, pages 1–5, September 2007.

[11] Z. Hasan, H. Boostanimehr, and V.K. Bhargava. Green cellular networks: A survey, some research issues and challenges. *IEEE Communications Surveys & Tutorials*, 13(4):524–540, 2011.

[12] H. Holma and A. Toskala. *WCDMA for UMTS: HSPA Evolution and LTE*, 5th edition. Wiley, September 2010.

[13] G.Y. Li, Z. Xu, C. Xiong, C. Yang, S. Zhang, Y. Chen, and S. Xu. Energy-efficient wireless communications: Tutorial, survey, and open issues. *IEEE Wireless Communications*, 18(6):28–35, December 2011.

[14] M. Nekovee. Cognitive radio access to TV white spaces: Spectrum opportunities, commercial applications and remaining technology challenges. In *Proceedings of IEEE Symposium on New Frontiers in Dynamic Spectrum*, pages 1–10, April 2010.

[15] T. Yucek and H. Arslan. A survey of spectrum sensing algorithms for cognitive radio applications. *Communications Surveys Tutorials, IEEE*, 11(1):116–130, 2009.

Biography

Tao Chen received his B.E. degree from Beijing University of Posts and Telecommunications, China in 1996, and Ph.D. degree from University of Trento, Italy in 2007, both in telecommunications engineering. Since 2008, he has been with VTT Technical Research Center of Finland, now working as senior researcher and project manager. From 2003 to 2007 he worked as a researcher in CREATE-NET, Italy. From 1996 to 2003, he was an engineer in a national research institute at China. His research interests include dynamic spectrum access, energy efficiency of heterogeneous wireless networks, cooperative communications, and wireless networking.

Marja Matinmikko received her M.Sc. degree in industrial engineering and management and Lic.Sc. degree in telecommunications from University of Oulu, Finland, in 2001 and 2007, respectively. She is currently finalizing her Ph.D. thesis on spectrum availability detection methods for cognitive radio systems (CRS). She joined VTT Technical Research Centre of Finland in Oulu, Finland, in 2001 and is currently working as senior scientist and project manager in CRS. She participated in the CEPT and ITU-R studies on spectrum demand calculation for IMT-Advanced in preparation for the World Radiocommunication Conference in 2007 (WRC-07) that led to the new spectrum identification for the mobile service. Her current research interests include technical, regulatory and techno-economic aspects of CRS. She is leading a national project consortium on cognitive radio trial environment (CORE) in the Finnish Trial program of Tekes - the Finnish Funding Agency for Technology and Innovation. She is management committee (MC) member of COST Action IC0905 TERRA where she is vice-chair of WG2 on CR/SDR Co-existence studies. She follows the spectrum regulatory activities on CRS at CEPT and ITU-R and contributes and participates in the CRS studies at ITU-R WP5A.

Javier (Javi) Del Ser was born in Barakaldo (Spain) in 1979. He obtained his combined B.S. and M.S. degree in Electrical Engineering from Faculty of Engineering (ETSI, www.ingeniaritza-bilbao.ehu.es) of the University of the Basque Country (Spain) in May 2003. In October 2006 he received his Ph.D. degree (cum laude) in Electrical Engineering, from Centro de Estudios e Investigaciones Tecnicas de Gipuzkoa (CEIT), San Sebastian (Spain). From 2003 to 2005 he was a teaching assistant at TECNUN (University of Navarra, www.tecnun.es). From August to December 2007 he was a visiting scholar at University of Delaware (USA), and from February to December 2008 he was an assistant professor and researcher at the University of Mondragon, Spain. Currently he is the technology manager at the OPTIMA department of TECNALIA RESEARCH & INNOVATION (www.tecnalia.com). His research interests are focused on communication and network information theory, iterative (Turbo) joint source-channel decoding and equalization, factor graphs, network coding, cognitive radio systems, wireless sensor networks, multimedia over wireless and heuristic optimization algorithms. He has also been granted twice with the Torres Quevedo contract grant from Spanish Ministerio de Ciencia e Innovacion (2007 & 2009), and is a senior member of the IEEE Communications, Signal Processing, Information Theory, Computational Intelligence and Photonics societies. Recently he has been awarded the "Talent of Bizkaia" prize for his outstanding professional curriculum.

Jianhua Zhang received her Ph.D. degree in circuit and system from Beijing University of Posts and Telecommunication (BUPT) in 2003. She is an associate professor from 2005 and now she is professor of BUPT. She has published more than 100 articles in referred journals and conferences. She was awarded the "2008 Best Paper" of *Journal of Communication and Network*. In 2009, she received a second prize award by CCSA for her contributions to ITU-R and 3GPP in IMT-Advanced channel model. In 2011, she was awarded the "New Century Excellent Talents in University" by MOE and "Young Talent Teachers" by Fok Ying Tung Education Foundation. Her current research interests include propagation models, green techniques for IMT-Advanced and beyond system, Relay and CoMP system, etc.

Energy Efficient Techniques for Resource Allocation in Cognitive Networks

E. Del Re[1,2], P. Piunti[1], R. Pucci[1,2] and L.S. Ronga[2]

[1]*Department of Electronics and Telecommunications Engineering,
University of Florence, 50100 Florence (FI), Italy;
e-mail: {enrico.delre, pierpaolo.piunti}@unifi.it*
[2]*CNIT (Florence Research Unit), 50136 Florence (FI), Italy;
e-mail: {renato.pucci, luca.ronga}@cnit.it*

Received 15 October 2011; Accepted: 11 May 2012

Abstract

The Cognitive Radio paradigm is aimed to optimize the utilization of licensed spectrum bands thanks to coexistence within the same network of licensed (primary) and cognitive (secondary) users. In this context, one of the most important key aspects is represented by an efficient resource allocation between secondary and primary users. Modeling it as an optimization problem, this paper provides a modified version of the well-known Iterated Water-Filling algorithm and a novel approach based on a game theory framework to solve this issue in a distributed and fair way. In particular, the proposed game is formulated as an S-Modular Game, since it provides useful tools for the definition of multi objective distributed algorithms in the context of radio communications. This paper provides also a performance comparison among the proposed solutions and the Simulated Annealing algorithm, that represents one of the most frequently used technique in this context.

Keywords: cognitive radio, resource allocation, game theory, energy efficiency.

Journal of Green Engineering, Vol. 2, 329–346.

1 Introduction

The radio spectrum efficiency represents nowadays a significant problem, due to the fast development of a large number of radio technologies in the last decade. Recent spectrum analysis points out that a large number of assigned spectrum bands are underutilized in either time domain or spatial domain [1]. In this context, Cognitive Radio (CR) [2] offers a smart paradigm aimed to optimize the utilization of the radio resource, allowing cognitive users to share the spectrum bands with licensed users. One of the most relevant open issues is represented by the identification of efficient methods to distribute and manage radio resources. In particular, taking into account the power consumption problem, the definition of energy efficient power allocation strategies could represent a key feature in designing cognitive networks.

In order to increase frequency utilization efficiency, in a cognitive radio network a dynamic spectrum access (DSA) [3] has been used. In particular, the term DSA covers several approaches to spectrum access that can be distinguished in three different models, on the basis of the categorization made by DYSPAN group in [4]: dynamic exclusive use model, open sharing model, hierarchical access model. In our approach we adopt a hierarchical spectrum access structure, identifying two kind of users: primary and secondary users. The licensed spectrum is assigned to the primary users (owners of the spectrum rights), while secondary (unlicensed) users can access spectrum following the *underlay approach*, transmitting below the noise floor of primary users. Thus, a transmitting secondary user represent an interference source both for primary and the other secondary users.

On the basis of the underlay approach, we refer to a cognitive radio network wherein secondary users are cognitive users, since they are intelligent and interact with selfish network users. Contrary to secondary users, primary users may be unaware of the presence of secondary users, even though they coexist within the same network and sharing the same frequency bands. Due to the necessity of frequent spectrum sensing and transmissions, secondary users may have strictly energy constraints, especially if they are battery powered in order to satisfy mobility requirements. A feasible strategy to reduce energy consumption is represented by the implementation of energy efficient techniques of power allocation.

In this scenario, a game theoretic framework allows us to study, model and analyze cognitive radio networks in a distributed way. Such an attractive feature allows us to achieve the flexibility and the efficient adaptation to the operative environment that were previously mentioned.

The paper is organized as follows: in Section 2 an overview on the application of game-theoretic approaches to spectrum sharing scenarios is illustrated, while in Section 3 the proposed system model and applicative scenario are presented. The game description and the Nash Equilibrium existence and uniqueness is discussed in Section 5, while in Section 4 the Water-Filling algorithm and a energy efficient modified version is reported. In Section 6 the results from computer simulation are commented. Finally some conclusions are expressed in Section 7.

2 Game Theory for Cognitive Radio

Due to the players' behavior, non-cooperative game theory is closely connected to mini/max optimization and typically results in the study of various equilibria, most notably the Nash equilibrium [5]. Developed cognitive radio strategy has been formulated according the mathematical discipline of Game Theory, with particular reference to S-Modular Games [6].

Non-cooperative games have been proposed for spectrum sharing in [7], which reports a detailed survey on game theoretic approaches for dynamic spectrum sharing in cognitive radio networks, by in-depth theoretic analysis and an overview of the most recent practical implementations. In [8], the authors investigate the issue about the spectrum sharing between a decentralized cognitive network and a primary system, comparing a suboptimal distributed non-cooperative game with the optimal solution power control algorithm and the method proposed in [9]. Apart from in the above-mentioned papers, the power control problem in spectrum sharing model is also discussed in [10, 11].

In [12, 13] the authors proposed different game-theoretic approaches to maximize energy efficiency of the users within wireless networks, making the utility functions being inversely proportional to the transmit power. Extending the above described results, this paper provides a distributed game-theoretic approach to obtain an energy efficient power allocation method that maximize the Signal to Interference-plus-Noise Ratio (SINR) level received by each user, taking into account throughput fairness among secondary users.

3 System Model

The proposed power allocation techniques refer to a cognitive radio context where a primary system (owner of the spectrum rights) coexisting with one

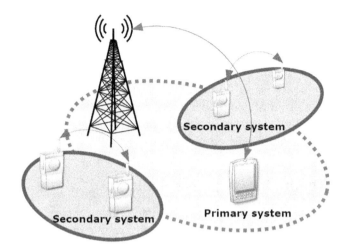

Figure 1 Example of coexistence between primary and secondary users.

or more secondary systems and sharing the same frequency band, as shown in Figure 1. Note that, considering a primary system in the network, the proposed scheme includes the possibility of existence of more than one primary user. Each secondary user is characterized by a dedicated sender and receiver, since each communicating couple consists of a transmitter site TX_i and a receiver site RX_i. Within each secondary transmission couple, we denote a *transmitter* and a *receiver* in order to identify the transmission course, however both the transmitter and receiver are assumed having a complete radio-frequency front-end and, therefore, both transmitter and receiver are able to transmit and receive data.

The system model uses a discrete-time model, based on iterations (which we ahead refer as t). Indeed, for every single iteration, all users act only once and until the next iteration they cannot do anything else. Each iteration represents a status update of the system; the real-time span modeled by each iteration will depend on the definition of the MAC layer.

By definition, primary users should not undergo a degradation of the required Quality of Service (QoS) due to the sharing of the same frequency band with secondary users. For this reason, a constraint on maximum transmission power level for secondary users must exist. This limitation can be obtained introducing the "Interference Cap" [14]; this parameter represents the total interference that the primary system willing to tolerate in order to not undergo a degradation of the required QoS. In case of cooperation existence between primary and secondary users, the Interference Cap can be send directly from primary users to secondary users on a shared channel. Otherwise, when primary users may be unaware of the presence of secondary users, in the proposed system there cannot be a cooperation among primary and secondary users; thus the Interference Cap will be fixed a priori, i.e. equal to noise floor of primary user. The presence of a chosen interference cap represents an upper bound of the total transmit power of the secondary users on the shared channel.

Each secondary user will choose the more suitable transmission power in order to achieve the best transmission quality, ensuring low interference to other secondary users. For simplicity of exposition, we will consider a fixed primary interference cap and therefore a fixed maximum transmission power for the secondary users; this assumption can be made without altering the validity of the system, since variations of this value are relatively slow compared with the time of convergence of the algorithm. In case of wireless networks with high primary mobility and/or more strict delay needs, a delay efficient approach should be followed, see [15].

4 Energy Efficient Iterative Water-Filling Algorithm

Among power allocation methods, actually Water-Filling [16] is one of the most frequently used algorithm. This algorithm bases on the idea that a vase can be filled by a quantity of water equal to the empty volume of the vase. It is well-known in the literature that a channel can be filled by an amount of power depending on the existing noise level. In order to maximize data-rate, power allocation in a multiuser scenario can follow the water filling principle. Due to the frequency band sharing, the increase of number of secondary users in the network equals to an increase of interference. Indeed for increasing number of users, secondary users experience higher levels of interference.

The Iterative Water Filling (IWF) algorithm is obtained basing on the above previous considerations; iteratively, each user calculates its transmission power level P_i^t following the water-filling principle until an equilibrium is reached. The equilibrium state is reached when the algorithm returns the same power allocation set for n consecutive times. The transmission power level is calculated as follows:

$$P_i^{(t)} = \max \left\{ 0, \left(P_{max}^{(t)} - \frac{P_i^{(t)} g_{ii}}{\gamma_i^{(t)}} \right) \right\}, \quad i = 1, \ldots, N \tag{1}$$

where $P_i^{(t)}$ is the power level assigned at the user i in the iteration t, P_{max} is maximum power that can be transmitted in the channel (the *water level*) and g_{ii} is the channel gain between transmitter TX_i and receiver RX_i. The factors $\gamma_i^{(t)}$ is the SINR at the receiver side at step t and it is calculated as follows:

$$\gamma_i^{(t)} = \frac{g_{ii} P_i^{(t)}}{\sum_{k \neq i} g_{ik} P_k^{(t-1)} + \sigma^2} \qquad (2)$$

where σ^2 is the power level of noise. If

$$\frac{P_i^{(t)} g_{ii}}{\gamma_i(t)} > P_{max}$$

the interference plus noise value is higher than maximum power that can be transmitted in the channel, then $P_i^{(t)} = 0$ is assigned to the user i.

Running IWF algorithm for low interference environments and/or limited number of users, we obtain good performance in terms of SINR received by secondary users. However for increasing values of interference, the algorithm get worst; indeed, users experiencing the best channel conditions will transmit at high power levels, while users experiencing bad channel conditions will receive high interference values and then they will be inactivated (i.e. when a receiver is close to another transmitter). For this reason, IWF can be considered to be unfair.

As it is, IWF is an energy inefficient algorithm, since it bases on the maximization of the total transmission power of each user in order to obtain the best SINR level. However, for a fixed target data-rate, we can identify a minimum target value for the SINR. In Figure 2 the SINR trends are reported for increasing values of the maximum transmission power for a different number of users. Taking into account such SINR trends, we propose the following energy efficient modified version of the algorithm, called Energy Efficient Iterative Water-Filling (EEIWF). It allows us to maintain the fixed data-rate, using the lowest total transmission power level. Each user updates P_{max} every iteration as follows:

$$P_{max}^{(t)} = \begin{cases} \frac{P_{max}^{(t-1)}}{k}, & \gamma^{(t)} \geq \gamma^{(t-1)} \\ P_{max}^{(t-1)}, & \gamma^{(t)} < \gamma^{(t-1)} \end{cases} \qquad (3)$$

where $k > 1$ represents the reduction factor and it controls the convergence speed of the algorithm. Note that for $k = 2$ the algorithm becomes the bisection method.

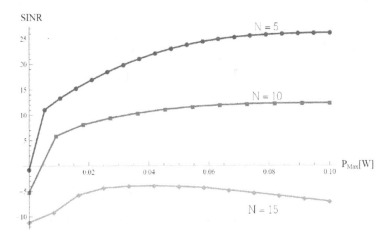

Figure 2 SINR trends for increasing values of maximum transmission power for different number of users.

The criterion expressed in (3) is enforced by observing that, running IWF, the SINR trend for increasing values of P_{max} is in the best case a monotonous increasing function and in the worst case a function with a maximum. The case of a monotonous decreasing function is not admitted by IWF. A continuous decreasing SINR when P_{max} increases means that an user experiments bad channel conditions: in this case user is inactivated.

EEIWF achieves the same SINR levels of IWF algorithm, allocating to secondary users an average transmission power level that is half of the one allocated by IWF; in Figure 3 are reported the average transmission power levels per user allocated by the two algorithm.

5 The Energy Efficient Fair Game

5.1 Game description

In this paper we propose a non-cooperative game with N secondary users, namely the players of the game, operating on one radio resource. This game can be easily extended considering a larger number of radio resources M (i.e. subcarriers of the same multi-carrier channel or different channels) following the approach proposed in [17], where subcarrier allocation is based on the normalized channel gain. Formally, the proposed non-cooperative game can be modeled as follows:

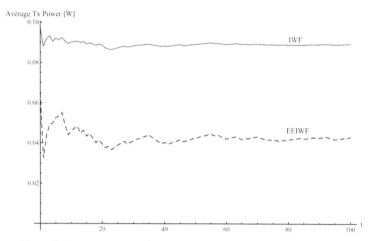

Figure 3 Average transmission power per user (Montecarlo simulation)

- Players: $\mathbf{N} = \{1, 2, 3, \ldots, N\}$ where $i \in \mathbf{N}$ is the i-th secondary user.
- Strategies: $\mathbf{S} = \{P_{\min}, \ldots, P_{\max}\}$.
- Utility Function: $u_i(p)$ where $i \in \mathbf{N}$ is the i-th secondary user and $p \in \mathbf{S}$ is the complete set of strategies.

We take into account the energy efficiency problem at the physical layer, considering an utility function expressed in bit/Joule as performance measure of the model [12, 18]. During the game each player tries to maximize the following utility function:

$$u_i(p(t), p(t-1)) = W \frac{R_i f(\gamma_i)}{p_i(t)} - \Omega_i(p(t), p(t-1)) p_i(t) \qquad (4)$$

where p is the complete set of strategies of all secondary users, W is the ratio between the number of information bits per packet and the number of bits per packet, R_i is the transmission rate of the ith user in bits/s, $f(\gamma_i)$ is the efficiency function (depending on the considered modulation), that represents a stochastic modeling of the number of bits that are successfully received for each unit of energy drained from the battery for the transmission, γ_i is the instantaneous SINR. Since the SINR depends on the path gains, each secondary user need to know them. In order to solve this problem that could have a strong impact on the signaling process, we assume that each receiver periodically send out a beacon, thanks to which transmitters can measure path gains. This procedure is feasible since both the transmitter and receiver are able to transmit and receive data, as presented before in Section 3. The period

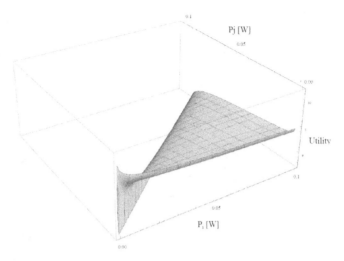

Figure 4 Trend of utility function (on z-axis) depending on transmission power levels (on x and y axes) for a two-player game; for each user in every single iteration the utility function have a maximum.

of beacon transmission should be chosen on the base of the coherence time of the channel.

In order to make the Nash Equilibrium of the game as efficient as possible (moving it closer to the Pareto Optimum), we consider the adaptive pricing function $\Omega_i(p_{i,-i})$ that generates pricing values basing on the interference generated by network users. Thus, the greater is the interference generated by a user transmitting at high power level, the greater will the value of pricing it will be pay, due to the fact that $\Omega(p)$ is strictly increasing with p.

The pricing function is written as follows [18]:

$$\Omega_i(p(t-1)) = \beta - \delta \exp\left(-\mu \frac{p_i(t-1)\sum_{i=1,k\neq i}^{N} g_{k,i}}{I_i^r(p_{-i}(t-1), P)}\right) \tag{5}$$

where p is the complete set of strategies of all secondary users, P is the power transmitted by the primary user and g_{ii} is the channel gain between transmitter TX_i and receiver RX_i. The term $I_i^r(p_{-i}(t-1), P)$ represents the total interference received by the ith user and it can be wrote as

$$I_i^r(p_{-i}(t-1), P) = \sum_{k\neq i} g_{k,i} p_k(t-1) + \sigma^2 + g_{12,i} P \tag{6}$$

Moreover, the pricing function bases on:

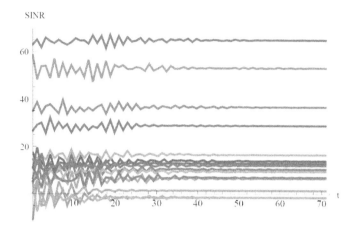

Figure 5 SINR convergence in a 15-user simulation with $\delta = 10^4$.

- $\beta > 1$ is the maximum pricing value,
- $\delta > 1$ is the price weight of the generated interference,
- $\mu > 0$ is the sensitivity of the users to interference.

These three parameters represent an useful tool to adapt the pricing function to the considered wireless network requirements, i.e. decreasing the value of δ the algorithm converges faster; we can force all secondary users to transmit at lower power levels increasing their sensitivity to the interference [18].

Thanks to the definition of the utility function as given in (4) and taking into account the pricing function given in (5), for each user in every single iteration the utility function have a maximum, depending on the transmission power levels of all the players in the game, as shown in Figure 4 for a two-player simulated game.

Simulation results show that the proposed algorithm has a fast convergence, also for large numbers of secondary users in the networks, as shown in Figure 5, where the SINR levels measured by each of the 15 users of the simulation are reported.

5.2 Existence and Uniqueness of the Nash Equilibrium

Under certain conditions, a Nash Equilibrium [19] offers a stable outcome and it can be guaranteed to exist, but does not necessarily mean the best payoff for all the players involved, especially in presence of pricing techniques. In the literature there are lots of mathematical methods to demonstrate the existence and uniqueness of Nash Equilibrium, like graphical [18, 20], quasi-concavity curve [21] and super-modularity [12].

Supermodular Games represent an interesting class of games since there are several compelling reasons like existence of pure strategy Nash Equilibrium, dominance resolvability, identical bounds on joint strategy space etc. that make them a strong candidate for resource allocation modeling. Supermodular games are based on the "supermodularity" concept, which is used in the social sciences to analyze how one agent's decision affects the incentives of others.

S-Games are normal form games $\Gamma = \langle N, S, \{f_i\} \rangle$ where N is the set of users, S the strategy space, f_i the set of utility functions and $\forall i \in N$ these conditions are satisfied:

1. the strategy space S_i of user i is a complete lattice.
2. f_i is supermodular in s_i.
3. f_i presents increasing differences in s.

The proposed utility function in (4) can be easily demonstrated to be supermodular, since:

1. the strategy space P is a complete lattice;
2.

$$\frac{\partial^2 u_i(p)}{\partial p_i \partial p_j} \geq 0 \tag{7}$$

for all $p \in P$ and $i \neq j$.
3. the utility function has the increasing difference property.

For details of the proofs we refer to [12], under the proposed conditions. Uniqueness of the Nash Equilibrium can be also demonstrated following the same approach, since we use a Best Response rule. Even if our proposed pricing function is more complicated, in comparison with the above-cited work, the demonstration procedure does not change. Indeed, the pricing function $\Omega(p(t-1))$ can be considered linear in $p(t-1)$, since the coefficient of $p(t-1)$ at time t is a constant.

6 Simulation Results and Performance Comparison

In order to evaluate the performance of a cognitive network based on the our proposed methods, we run Montecarlo simulations; in this paragraph we show the obtained results. The operating context is a terrain square area of 1 km edge, with a suburban path-loss profile. Primary transmitter and receiver positions are fixed; secondary transmitters are independently located in the area, while the secondary receivers positions are placed randomly in a 200 m diameter circle around the respective transmitters. Each secondary user transmits isotropically with $p_i \leq P_{max}$, where $P_{max} = 20$ dBm on the base of a fixed interference cap. Moreover, we consider a noise power $\sigma^2 = -100$ dBm, frequency $f = 1GHz$, $W = 4/5$, a common rate $R_i = 10$ kbit/s, $\beta = 10^4$, $\delta = 10^4$ and $\mu = 10^{-2}$; the values of the pricing parameters are chosen in order to obtain a fast convergence of the algorithm (depending on $\beta - \delta$) and to set a low sensitivity of the users toward interference (depending on μ). In order to obtain a qualitative evaluation of the proposed power allocation methods, we decide to compare their performance with an optimal centralized heuristic power allocation system, like Simulated Annealing (SA) [22]. The mean value of the SINR received by secondary users has been chosen as the performance index for the three optimization methods. We run Montecarlo simulations for increasing number of secondary users, while all the other parameters of the system remain the same of the previous mentioned operating context.

Simulation results illustrated in Figure 6 show clearly that the SA and the proposed game have similar performance: the curves have the same trend, but the game's one is on average 1.5 dB below. On the other hand, Water-Filling obtains lower mean SINR levels and performance worsens for increasing number of users in the network, as highlighted previously.

In addition to the SINR, the energy efficiency of the three considered methods is an another important key feature that we need to investigate. If the SINR performance are quite the same for the proposed game and the SA, on the contrary we can observe a great difference in terms of power allocations. Indeed, Figure 7 shows an example of mean transmission power levels allocated by the three investigated methods at the end of a simulation with 15 users; in this figure we can observe that SA allocation uses more power than game allocation to obtain similar SINR performance, as shown in Figure 6. For what concern the EEIWF, while some users are switched off, the others transmit at highest levels, compared with the other two proposed allocations. In Figure 7 the power allocation of the proposed game is shown in purple, in yellow is reported the additional power allocated by SA (with respect to game) and in blue the excess additional power allocated by EEIWF (with respect to SA).

Looking at the results from a general network view, we can easily observe that the power allocation obtained by the proposed game is the most fair, since all the users are able to transmit, even if they are experiencing very bad

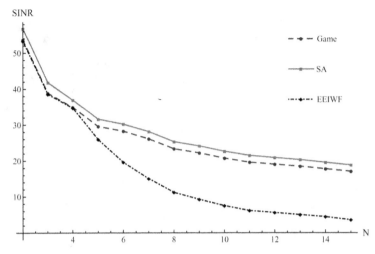

Figure 6 Trends of SINR mean values for increasing number of secondary users in the network.

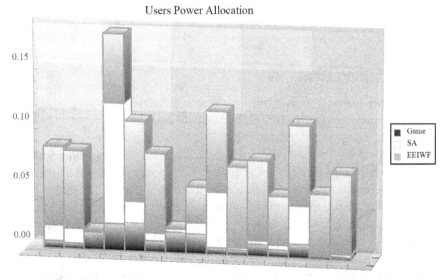

Figure 7 Example of mean values of power allocation for a 15 users network obtained thanks to Monte-Carlo method.

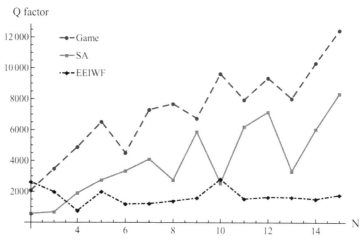

Figure 8 Trend of Q for increasing number of users.

channel condition. In order to obtain a qualitative estimation that takes into account at the same time SINR performance and energy efficiency of the three considered methods, in Figure 8 is reported the trend Q, that represents the mean value of the ratio between the SINR level received and allocated power of the transmitter, calculated for each user and obtained thanks to Monte-Carlo method.

7 Concluding Remarks

In this paper we provide two different power allocation methods, a modified version of the Iterated Water-Filling, called EEIWF, and a game theoretic framework based on S-Modular game. Both these methods take into account energy efficiency in a cognitive network, wherein primary and secondary users coexist. Transmission power of secondary users is limited by the presence of an interference cap, defined as the total interference that primary users willing to tolerate, without loosing their required QoS. Moreover, in the proposed game secondary users are discouraged to transmit at high power levels, since they are charged on the base of the interference they generate, thanks to the introduction of a pricing function inside of the utility function. Simulation results show a fast convergence of the proposed game also for a large number of users in the cognitive network. Moreover, the EEIWF is demonstrated to achieve the same SINR levels of IWF algorithm, allocating to secondary users an average transmission power level that is half of the one allocated by IWF.

A performance comparison among the proposed game, an optimal centralized resource allocation method (Simulated Annealing) and the EEIWF is also included. Simulation results show clearly that the proposed game converges to the same SINR values obtained from the heuristic optimization method and, in general, game theory obtains better performance than EEIWF. Moreover, the proposed game results to be the most energy efficient, also for a large number of considered users. Further investigations will be made in order to quantify and analyze the signaling process among secondary users.

References

[1] G. Staple and K. Werbach. The end of spectrum scarcity [spectrum allocation and utilization]. *Spectrum, IEEE*, 41:48–52, March 2004.

[2] S. Haykin. Cognitive radio: Brain-empowered wireless communications. *IEEE Journal on Selected Areas of Communications*, 23(2):201–220, February 2005.

[3] D. Hatfield and P.J. Weiser. Property rights in spectrum: Taking the next step. In *Proceedings of the First IEEE International Symposium on New Frontiers in Dynamic Spectrum Access Networks (DySPAN2005)*, pages 43–55, November 2005.

[4] IEEE std 1900.1-2008. 1900.1-2008 – IEEE Standard Definitions and Concepts for Dynamic Spectrum Access: Terminology Relating to Emerging Wireless Networks, System Functionality, and Spectrum Management. IEEE Standards Association, July 2011.

[5] L.S. Ronga, R. Pucci, and E. Del Re. Energy efficient non-cooperative methods for resource allocation in cognitive radio networks. *Communications and Networks*, 4:1–7, 2012.

[6] D.M. Topkis. Equilibrium points in nonzero-sum *n*-person submodular games. *SIAM J. Control Optim.*, 17:773–778, October 1978.

[7] B. Wang, Y. Wu, and K.J. Ray Liu. Game theory for cognitive radio networks: An overview. *Computer Networks*, 54:2537-2561, April 2010.

[8] R. Luo and Z. Yan. Power allocation using non-cooperative game theoretic analysis in cognitive radio networks. In *Proceedings of WiCOM*, September 2010.

[9] D. Li, X. Dai, and H. Zhang. Game theoretic analysis of joint rate and power allocation in cognitive radio networks. *International Journal on Communications, Network and System Sciences*, 1:1–89, February 2009.

[10] G. Bansal, Md.J. Hossian, and V.K. Bhargava. Adaptive power loading for OFDM-based cognitive radio systems. In *Proceedings of IEEE Conference on Communications*, pages 5137–5142, August 2007.

[11] J. Huang, R. Berry, and M.L. Honig. Auction-based spectrum sharing. *ACM/Springer Journal of Mobile Networks and Applications (MONET)* (Special issue on WiOpt'04), 11(3):405–418, June 2006.

[12] C.U. Saraydar, N.B. Mandayam, and D. Goodman. Efficient power control via pricing in wireless data networks. *IEEE Transaction on Communications*, 50:291–303, February 2002.

[13] R.D. Yates. A framework for uplink power control in cellular radio systems. *IEEE Journal on Selected Area of Communications*, 13, September 1995.

[14] L. Tianming and S.K. Jayaweera. A novel primary-secondary user power control game for cognitive radios with linear receivers. In *Proceedings of IEEE Military Communications Conference (MILCOM 2008)*, November 2008.

[15] F. Meshkati, H.V. Poor, and S.C. Schwartz. Energy efficiency-delay tradeoffs in CDMA networks: A game theoretic approach. *IEEE Transactions on Information Theory*, 55(7):3220–3228, July 2009.

[16] S. Deng, T. Weber, and A. Ahrens. Capacity optimizing power allocation in interference channels. *AEU – International Journal of Electronics and Communications*, 63(2):139–147, January 2008.

[17] D. Wu, D. Yu, and Y. Cai. Subcarrier and power allocation in uplink OFDMA systems based on game theory. In *Proceedings of IEEE International Conference on Neural Networks & Signal Processing*, Zhenjiang, China, June 2008.

[18] C.K. Tan, M.L. Sim, and T.C. Chuah. Fair power control for wireless ad hoc networks using game theory with pricing scheme. *IET Communications*, 4(3):322-333, November 2008.

[19] D. Fudenberg and J. Tirole. *Game Theory*. MIT Press, Cambridge, MA, August 1991.

[20] F. Nan, M. Siun-chuon, and N.B. Mandayam. Pricing and power control for joint network-centric and user-centric radio resource management. *IEEE Transactions on Communications*, 52:1547–1557, September 2004.

[21] H. Zhu, Z. Ji, and K.J.R. Liu, 'Fair multiuser channel allocation for OFDMA networks using Nash bargaining solutions and coalitions. *IEEE Transactions on Communications*, 53:1366–1376, August 2005.

[22] S. Kirkpatrick et al. Optimization by simulated annealing. *Science*, 220(4598):671-680, May 1983.

Biography

Enrico Del Re was born in Florence, Italy. He received the Dr. Ing. degree in electronics engineering from the University of Pisa, Pisa, Italy, in 1971. Until 1975 he was engaged in public administration and private firms, involved in the analysis and design of the telecommunication and air traffic control equipment and space systems. Since 1975 he has been with the Department of Electronics Engineering of the University of Florence, Florence, Italy, first as a Research Assistant, then as an Associate Professor, and since 1986 as Professor. During the academic year 1987-1988 he was on leave from the University of Florence for a nine-month period of research at the European Space Research and Technology Center of the European Space Agency, The Netherlands. His main research interest are digital signal processing, mobile and satellite communications, on which he has published more than 300 papers, in international journals and conferences. He is co-editor of the book *Satellite Integrated Communications Networks* (North-Holland, 1988), one of the authors of the book *Data Compression and Error Control Techniques with Applications* (Academic, 1985) and editor of the books *Mobile and Personal Communications* (Elsevier, 1995), *Software Radio Technologies and Services* (Springer, 2001), *Satellite Personal Communications for Future-Generation Systems* (Springer, 2002), *Mobile and Personal Satellite Communications 5-EMPS2002* (IIC, Italy, 2002) and *Satellite Communications and Navigation Systems* (Springer, 2008). He has been Chairman of the European Project COST 227 "Integrated Space/Terrestrial Mobile Networks" (1992–1995) and the EU COST Action 252 "Evolution of satellite personal communications from second to future generation systems" (1996–2000). He has been Chairman of the *Software Radio Technologies and Services Workshop* (2000), the *Fifth European Workshop on Mobile/Personal Satcoms* (2002) and the *Satellite Navigation and Communications Systems* (2006). He received the 1988/89 premium from the IEE (UK) for his paper entitled "Multicarrier demodulator for digital satellite communication systems". He is Head of the Signal Processing and Communications Laboratory of the Department of Electronics and Telecommunications of the University of Florence. Presently he is President of the Italian Interuniversity Consortium for Telecommunications (CNIT), having served before as Director. Professor Del Re is a Senior Member of the IEEE and a member of the European Association for Signal Processing (EURASIP).

Pierpaolo Piunti received the Degree in Telecommunications Engineering from University of Florence on April 2011. From June 2011 to December 2011 he worked as contractor ad Interuniversitary Italian Consortium for Telecommunications (CNIT). Actually he is a Ph.D. student of University of Florence. His research activity is focused on energy efficiency of communications networks and cognitive radio.

Renato Pucci received his degree in Telecommunication Engineer From the University of Florence in 2008, winning with his thesis the AICT & AEIT awards as *Best 2008 Italian Thesis on ICT scientific innovation* and as *Youngest researcher in ICT scientific innovation* and being one of the four finalists of the *Best italian Thesis of 2008* AICA award. In 2012 Renato received his Ph.D. from the University of Florence, collaborating with Thales Alenia

Space SpA on the evaluation of applicability of Wireless Sensor Networks in spatial exploration context.

In 2012 he joined the Italian National Consortium for Telecommunications (CNIT). His areas of interest are Cognitive Radio, Wireless Communications, Game Theory and Telecommunication in spatial context.

Luca Simone Ronga [IEEE S89-M94-SM04] received his M.S. degree in electronic engineering in 1994 and his Ph.D. degree in telecommunications in 1998 from the University of Florence, Italy. In 1997 joined the International Computer Science Institute of Berkeley, California, as a visiting scientist. In 1998 obtained a post-doc position in the engineering faculty of the University of Florence. In 1999 he joined the Italian National Consortium for Telecommunications, where he is currently head of research. He conducts research activity and project management in various telecommunications areas, mainly in the satellite and terrestrial wireless fields. He has been leader of national and international research groups. He authored over 50 papers published in international journals and conference proceedings. He has been editor of *EURASIP Newsletter* for four years. His interests range from satellite communications to Software Defined Radio and Cognitive Radio techniques.

Energy-Efficiency of Cooperative Sensing Schemes in Ad Hoc WLAN Cognitive Radios

Reshma Syeda and Vinod Namboodiri

*Department of Electrical Engineering and Computer Science,
Wichita State University, Wichita, KS 26260, USA;
e-mail: {rssyeda, vinod.namboodiri}@wichita.edu*

Received 21 February 2012; Accepted 11 May 2012

Abstract

In cognitive radio networks, the secondary users need to coordinate among themselves to reap the benefits of cooperative spectrum sensing. In this paper, we study and analyse the energy efficiency of two generic cooperative sensing schemes in an ad hoc WLAN backdrop – distributed cooperative sensing scheme and centralized cluster based cooperative sensing scheme. We further propose corresponding enhanced and adaptive versions of these two schemes where only a fraction of nodes sense in each sensing cycle, as opposed to all the nodes in the network. Using an analytical energy model for sensing, we quantify the energy costs of each of these schemes and perform a comparative numerical analysis to demonstrate the amount of energy savings of the proposed cooperative schemes over their generic counterparts and non-cooperative schemes.

Keywords: cognitive radio, cooperative sensing, energy efficiency, WLAN.

1 Introduction

The world has witnessed a great deal of change this decade when it comes to wireless technologies. Be it Wi-Fi, WiMAX, WSNs or any other subtechnology belonging to one of these areas, these technologies have permeated so deep into our lives that it is hard to imagine our lifestyles without them. However, with demand comes scarcity. The spectrum allocated to these technologies is greatly limited and hence the scarcity which emphasizes the great need to deal with and overcome this issue. The spectrum measurements taken by BWRC and the shared spectrum company clearly show that only a few frequency bands are in use while about 70% of the remaining spectrum (primarily belonging to legacy radio technologies) remains unused for longer periods of time [1, 2]. This is where the opportunistic spectrum access (OSA) comes into picture. Through OSA, users can utilize the spectrum currently being unused especially when it comes to licensed spectrum of legacy technologies. One example is the TV spectrum. However, for a radio to use the spectrum opportunistically it should be aware and intelligent enough to look out for such vacant spectrum bands. Hence there is a great need to build better and intelligent radios and this is where the phrase 'cognitive radio' comes from. Cognitive Radios (CRs) opportunistically cash in on the licensed spectrum allocated to rightful owners that is not being used in time, frequency, space and code dimensions of a signal at a given instant (also called spectral opportunities) in order to make their communications efficient in terms of throughput, energy and delay metrics. Lately this term has become synonymous with the term Dynamic Spectrum Access (DSA) in the sense that the goal of DSA/OSA is achieved through the cognition of the radio.

Dealing with the Wi-Fi crowding phenomenon is critically important for sustainable wireless computing for our future generations. One of the most recent endeavors in this direction was made by FCC which approved and reallocated the use of Television White Spaces (TVWS) for Wi-Fi. Such renovated Wi-Fi technology capable of using TVWS via cognition was renamed as White-Fi (IEEE standard 802.11af), sometimes also called 'Wi-Fi on Steroids'. TV channels 2–52 have been opened up for unlicensed usage by the general public. Though the FCC recently has agreed to ditch the spectrum sensing requirement and instead encourage the use of online databases for spectrum vacancy [22, 23], it would still be an undeniable component when considering an 'on the fly' ad hoc Wi-Fi network that does not have access to these online databases.

In the field of CR technology, the rightful users of the licensed spectrum are termed the Primary Users (PUs) whereas the other CR users trying to use this spectrum opportunistically are the Secondary Users (SUs). The SUs, before dynamically accessing this licensed spectrum should make sure that it is not being used by any PU in their local vicinity so as to avoid interference to the PUs. The key component to achieve this is the sensing of the spectrum with high reliability.

As the transmitter based sensing techniques like energy detection [4, 5], matched filtering [6], cyclostationary [7] feature detection rely solely on the PU signal detection, there is a high chance for the cognitive user to be blinded by fading, shadowing and interference which might further degrade the accuracy of these techniques. The effects of these degradations have been studied in [8]. To overcome these blinding phenomena, cooperative sensing has been found to be far superior to just local/non-cooperative sensing as it solves the hidden node problem in addition to the others mentioned.

Based on the architecture, cooperative sensing schemes can be categorized into *distributed* and *centralized* cluster based schemes.

In a distributed cooperative sensing scheme, the SUs share the sensed information among themselves and make individual decisions. The advantage here is that there is no need for a common receiver infrastructure and high bandwidths [10]. Distributed collaboration schemes are discussed in [11]. In [12, 13] Relay based cooperation is discussed where a few SUs act as relays to other SUs. In [13] the Amplify and Relay (AR), Detect and Relay (DR) schemes are proposed for sensor networks.

In a centralized cooperative sensing system model, there is a common receiver which collects all the sensing information done by the SUs (CRs). Our interest is mainly on the cluster based schemes where the SUs are grouped into clusters or teams for collaboration; the reason being that to design an energy efficient cooperative sensing scheme, grouping would be a key component to cash in on the team collaboration instead of placing the burden of sensing a very large spectrum on each SU [14]. By grouping the SUs into clusters and selecting the most favourable SU in every cluster to report to the common receiver, the sensing performance can be greatly enhanced [15]. Guo and Peng [16] studied the optimal number of clusters required to minimize the communication overhead without loss in the detection performance. In [17], a two level hierarchical cluster based architecture is proposed where the low level collaboration is among the SUs within a cluster and high level collaboration is among the cluster heads chosen for each cluster. Other grouping techniques are studied in [18–21].

Most of the work done till to date on cooperative sensing in CR networks mainly focuses on increasing the throughput and spectrum sensing efficiency/accuracy. However, there is limited literature that looks at the energy cost aspect of the cognitive radio technology. One work that does and is our point of interest is [9] which looks at the positive and negative impact of CR MAC layer sensing in terms of energy. However, this work does not look at the energy costs of cooperative sensing; and the techniques (optimal scan and greedy scan schemes) proposed here are non-cooperative sensing schemes. Our work in this paper is a subsequence to that, with a goal to minimize the energy spent in spectrum sensing for each node and the network as a whole through cooperative sensing schemes. Also, there are many variations of cooperative sensing schemes with no consensus over which specific scanning scheme performs the best. Hence, we chose to look at a generic class of distributed and centralized schemes and apply our energy model to evaluate their energy consumption. We also propose the improved versions of both the generic distributed and centralized scheme and demonstrate the relative energy savings through evaluations. The following are the specific contributions made in this paper:

- We developed an energy model to study and quantify the energy costs of a broader class of existing generic cooperative MAC layer sensing schemes – one based on a distributed architecture and one based on a centralized architecture.
- We proposed corresponding new α-schemes: α-distributed and α-centralized, where only a fraction α of the nodes scan as opposed to all the nodes in the generic distributed and centralized schemes.
- We numerically evaluated and compared the energy consumption costs of the generic distributed and centralized schemes against the proposed α-distributed and α-centralized schemes.
- We studied optimal values of α and number of nodes in the network N.
- We made a conclusive energy comparison study of non-cooperative sensing schemes and cooperative schemes.

The rest of the paper is organized as follows: Section 2 gives an overview of the current sensing schemes and the proposed α-schemes along with the system model. Further energy consumption analysis equations are developed in Section 3 and energy savings analysis is done. Numerical evaluations of these equations is carried out in Section 4, followed by an interpretation of their significance.

2 Overview of Sensing Schemes

Spectrum sensing is a key critical component in the cognitive radio technology since the ability of cognition is achieved through this spectrum sensing

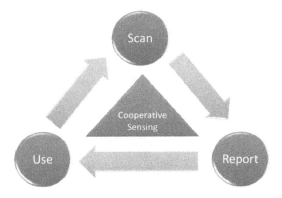

Figure 1   Cooperative sensing cycle.

functionality of the CR. These spectrum sensing schemes can be roughly classified into two categories: non-cooperative sensing and cooperative sensing.

2.1 Non-Cooperative Sensing

Non-cooperative sensing schemes also called as local sensing schemes require each node in the network to sense the spectrum for free channels individually. These nodes do not share the scanned information with their neighbours and hence no reporting is involved in a non-cooperative sensing cycle. The non-cooperative sensing we use as reference in our work is the optimal scan scheme from [9]. In this optimal scan scheme, every node has to scan all the channels before choosing the optimal channel among them.

2.2 Cooperative Sensing

In cooperative sensing schemes, all the nodes in the network sense/scan the spectrum and share this information with all the neighbouring nodes in the reporting phase (see Figure 1). This process reduces the CR blinding due to interference, fading and hence increases the probability of PU detection. To be precise, the collaboration of SUs can be used to either increase the number of channels scanned or improve the detection probabilities by having multiple nodes scan a single channel.

In this paper, we discuss more of the parallel cooperative sensing [24] where all the nodes are designated different channels to be scanned. Each node individually scans the designated channels in parallel (concurrently) and

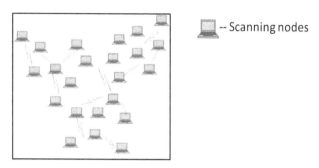

Figure 2 Distributed scheme.

then conveys this information in the form of sensing reports (SR). Each SR is of a constant size with an SR field of fixed bit map based on the total number of channels to be scanned by the network of nodes.[1]

Each sensing cycle (SC) has a scanning period (SP) and a reporting period (RP). Based on how the reporting/sharing is accomplished in the reporting period, cooperative sensing can be categorized into mainly two broad classes of architectures – distributed and centralized. The generic distributed sensing scheme we refer in this work belongs to the distributed architecture and is hereafter referred to as 'distributed sensing scheme' and the generic centralized cluster based scheme belongs to the centralized architecture which is hereafter referred to as the 'centralized cluster based scheme'.

2.2.1 Distributed Sensing Scheme

In this distributed sensing scheme, each node scans its share of channels during the SP and shares this information with all its neighbors in the RP (see Figure 2).

2.2.2 Centralized Cluster Based Sensing Scheme

In this centralized cluster based scheme (see Figure 3), the whole network of nodes can be divided into clusters (based on some higher layer protocol). Each cluster has a cluster head (CH) which does not carry out the scanning. Only the cluster members (CMs) in each cluster scan their share of the channels and convey this information to their respective CH. The CHs then share this information with the other remaining CHs in the network. There are two

---

[1] A fixed size of SR simplifies protocol design and allows dynamic re-configurations of protocol parameters (like the number of nodes that will scan the spectrum) without impacting other protocol fields.

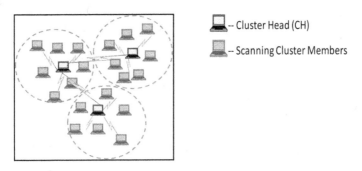

Figure 3 Centralized scheme.

levels of communication happening here, one at the intra cluster level and the other at the inter cluster level between the CHs. At the end of the inter cluster level information exchange, all the nodes in the network have a global view of the channel information for the whole network.

2.3 Proposed α-Sensing Schemes

Next, we propose the α-sensing schemes to make the conventional schemes more energy efficient. In the α-sensing schemes only a fraction of nodes share the burden of scanning the channels in each sensing cycle. This is a form of load sharing which saves energy for the other fraction of nodes in that sensing cycle and simultaneously the scanning of all the channels is collectively completed by the end of the sensing cycle. The α-sensing approach provides a 'smooth' way to handle and evaluate reduction in number of sensing nodes as opposed to schemes that some fixed number of nodes as sensing agents. These schemes are practical to implement with the fraction of nodes chosen based on their SNRs, PDRs (Probability Detection Ratios), shadow correlation, energy remaining, etc., to attain better spectrum sensing accuracies. The level of correlation among nodes could also be used to choose a value of α; a higher value of α can be used when expected correlation is low and vice-versa. We will concern ourselves more with the impact of a certain value of α in the rest of the paper leaving the selection of α for future work. The following subsections describe two types of α-sensing and how they could be implemented practically.

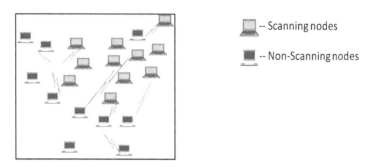

Figure 4 α-distributed scheme.

2.3.1 α-Distributed Sensing Scheme

In the α-distributed sensing scheme (see Figure 4), only an α fraction of nodes scan the channels in a given sensing cycle. However they still share this information with all their neighbouring nodes just similar to the generic distributed sensing scheme. This saves energy, as each node has to scan only an α percentage of the times on an average, in any given number of sensing cycles. This subset of α nodes can be chosen based on a probabilistic random number generator. For every SP, each node, through a probabilistic method (random number generator) decides to participate in scanning if the generated value is less than the α value. This way on an average each node in the network scans once in $(1/\alpha)$ sensing cycles.

2.3.2 α-Centralized Cluster Based Sensing Scheme

In this α-centralized sensing scheme (see Figure 5), instead of all the CMs in the cluster scanning for channels, only a fraction α of the CMs share the scanning responsibility and report this information to the CH. The CH in turn shares this information with other CHs similar to the centralized scheme. Once the CHs receive all the other cluster scan information, each CH shares a final report with its CMs. The fraction α of the CMs in each cluster is chosen by the CH.

Though another option of choosing a fraction α of the clusters instead of fraction of CMs from each cluster is possible, keeping in consideration the spatial diversity benefits of the clusters in the case of varying channel characteristics, it is better to have the scanning carried out in all the clusters for better sensing accuracies. To further study these schemes, the system model and the analytical energy model are developed in the next section.

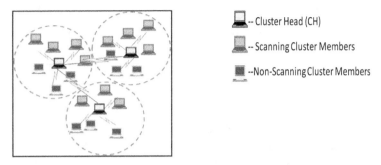

Figure 5 α-centralized scheme.

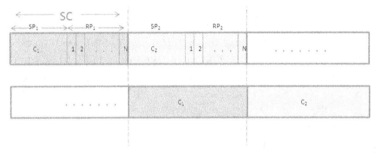

Figure 6 Time frame structures of scanning and reporting, data transmission respectively.

2.4 Basic Common Aspects of the Cooperative Sensing Model

We envision a static crowded ad hoc WLAN scenario where all the nodes are connected in a clique network and hence are in the hearing range of each other. Each node has two transceivers/radios – one completely dedicated for the sensing and reporting purpose and the other radio for data transmission (see Figure 6). The radio for sensing and reporting shifts to the channel(s) that needs to be scanned during the SP and finally switches to the control channel for reporting purposes which is shown in the first time frame structure. The other physical radio for data transmission switches to the vacant channel over which it could transfer the data after the first sensing cycle as shown in the second time frame structure. We assume a perfect channel scheduling.

At the end of the SP, each node shares a sensing report with all the remaining nodes through a single broadcast packet regardless of the number of channels scanned. Since the nodes perform parallel scanning, at the end of the reporting period every node would have the spectrum map of all the channels. We do not presume the existence of a fusion center in this 'on the fly' network for both the distributed and centralized architectures. In the distributed scheme, each node acts as a fusion center for itself while in the centralized scheme the cluster head can take up this role for its cluster. Section 5 later discusses the impact of changes to the assumed model on our results.

2.5 Orchestration

The total number of channels to be sensed C by the network of N nodes are decided prior to the start of SC. Since there is no reporting involved in the non-cooperative sensing scheme, each node scans all the C channels in the scanning period.

In the distributed cooperative sensing scheme, each node scans its share of designated channels in the SP. In the RP, it broadcasts this information to its $(N-1)$ neighbors and similarly decodes the $(N-1)$ reports it receives from its scanning neighbors. In the centralized scheme, each CM scans its share of designated channels and sends an SR to the CH. The CH after collecting all the SRs of its CMs, broadcasts an SR with this scan information of its cluster to all the other CHs in the network. Similarly, after receiving all the SRs from its peer CHs, each CH then sends a final consolidated SR having all the channel scan information to its CMs. Once these final SRs from the corresponding CHs are shared, all the CMs and CHs have a global view of all the channels scanned by the whole network.

2.6 Energy Model of Sensing Cycle

Energy consumed by each node per sensing cycle E_S is given by the sum of the total energy to scan all the assigned channel $E_{T\,scan}$, the total energy to switch between these channels ETsw, the total energy to transmit/broadcast the SR(s) to the other nodes E_{Tx} and the total energy to receive the SRs from all the other nodes E_{Rx}.

$$E_S = E_{T\,scan} + E_{T\,sw} + E_{Tx} + E_{Rx}$$

Based on the above system and energy model developed, appropriate equations for energy consumption analysis are derived in the following section.

Table 1 Definition of variables used [9].

Variable	Definition
P_{scan}	Power consumed to scan a channel (700 mW)
P_{sw}	Power consumed to switch once between channels (750 mW)
P_{tx}	Power consumed to transmit a Packet (750 mW)
P_{rx}	Power consumed to receive a Packet (750 mW)
T_{scan}	Time to scan a channel (50 ms)
T_{sw}	Time to switch once between channels (0.06 ms)
T_{data}	Time to transmit a Data packet (0.08 ms)
E_{scan}	Energy consumed to scan a channel $= P_{scan}T_{scan}$
E_{sw}	Energy consumed to switch once between channels $= P_{sw}T_{sw}$
E_{srt}	Energy consumed to transmit a Sensing Report (SR) $= P_{tx}T_{data}$
E_{srd}	Energy consumed to decode/receive a SR $= P_{rx}T_{data}$
C	Number of channels
N	Number of nodes in the network
N_S	Number of scanning nodes in the network
α	Fraction factor
K	Number of clusters into which the whole network of N nodes is divided/grouped
M	Number of CMs in the cluster
M_S	Number of scanning CMs in the cluster

3 Energy Consumption Analysis of Sensing Schemes

3.1 Non-Cooperative Sensing Scheme

Since there is no reporting period in a non-cooperative sensing scheme, the energy consumed by each node is just the sum of the energy to scan the designated channels and the energy consumed to switch between those channels. Each node scans channels and hence has to switch $(C-1)$ times when hopping from one channel to another. Hence

$$E_{non-coop}^S = CE_{scan} + (C-1)E_{sw}$$

where $E_{non-coop}^S$ is the energy consumed for each node during the scanning period. Total energy consumed by all the N nodes in the network to sense C channels in this non-cooperative sensing scheme is given by

$$E_{non-coop} = NE_{non-coop}^S$$
$$E_{non-coop} = NCE_{scan} + N(C-1)E_{sw} \quad (1)$$

3.2 Distributed Sensing Scheme

In this generic scheme, each node scans $C/N \ (\geq 1)$ channels in each SP as long as $N > 1$. If $N > C$, not every node has to scan the channels. Hence the total energy for the whole network in the distributed scheme is given by the following generic equation:

$$E_{dist} = N_S E_{dist}^S + (N - N_S)E_{dist}^{NS} \tag{2}$$

where $N_S = \min(N, C)$ is the number of scanning nodes, E_{dist}^S is the energy consumed by the scanning node, E_{dist}^{NS} is the energy consumed by the non-scanning node.

E_{dist}^S for a scanning node is the sum of energy to scan the designated (C/N_s) channels, switch in between those channels, transmit one SR with the channel scan information and receive $(N_s - 1)$ SRs from the other scanning nodes. E_{dist}^{NS} for a non-scanning node is the energy to decode all the SRs received from the scanning nodes.

$$E_{dist}^S = \left(\frac{C}{N_s}\right) E_{scan} + \left(\frac{C}{N_s} - 1\right) E_{sw} + E_{srt} + (N_s - 1)E_{srd} \tag{3}$$

$$E_{dist}^{NS} = N_s E_{srd} \tag{4}$$

Substituting equations (3) and (4) in (2) gives rise to the following energy equation for the whole network of nodes of distributed scheme

$$E_{dist} = C E_{scan} + (C - N_S)E_{sw} + N_S E_{srt} + N_S(N - 1)E_{srd} \tag{5}$$

Since $N_S = \min(N, C)$ we have the following two cases:

If $N \leq C$, then

$$E_{dist} = C E_{scan} + (C - N)E_{sw} + N E_{srt} + (N^2 - N)E_{srd} \tag{6}$$

If $N > C$, then only C nodes are required to scan

$$E_{dist} = C E_{scan} + C E_{srt} + (NC - C)E_{srd} \tag{7}$$

3.3 α-Distributed Sensing Scheme

In this scheme, for a chosen value of α ($0 < \alpha \leq 1$), only αN nodes perform the scanning while the remaining nodes do not scan for that SP. Hence for a given number of sensing cycles, the nodes in this scheme would have to scan only for an α percentage of the cycles on an average while they can save on their energy for the remaining $(1 - \alpha)$ percentage of the cycles.

Practically, since the sensing nodes αN cannot be either less than 1 or for that matter greater than C, we arrive at the following bound for

$$\left(\frac{1}{N}\right) \le \alpha \le \min\left(\frac{C}{N}, 1\right) \tag{8}$$

Because of this α bound we can completely rule out the possibility of $\alpha N > C$ in this α-scheme.

Substituting $N_S = \alpha N$ in Equations (2–4) gives the following energy equations for each of the scanning node and for the whole network respectively:

$$E^S_{\alpha-dist} = \left(\frac{C}{\alpha N}\right)E_{scan} + \left(\frac{C}{\alpha N} - 1\right)E_{sw} + E_{srt} + (\alpha N - 1)E_{srd} \tag{9}$$

$$E_{\alpha-dist} = CE_{scan} + (C - \alpha N)E_{sw} + \alpha N E_{srt} + \alpha(N^2 - N)E_{srd} \tag{10}$$

where $E^S_{\alpha-dist}$ is the energy consumed by the scanning node and $E_{\alpha-dist}$ is the energy consumed by the whole network of nodes.

3.4 Centralized Cluster Based Sensing Scheme

In this generic centralized scheme, the network of N nodes is divided into K clusters, each cluster having a group of $M+1$ nodes such that $N = K(M+1)$. Each cluster of $M + 1$ nodes thus has one cluster head (CH) and M cluster members (CMs). The CMs alone do the scanning and send their SRs to their respective CHs. Each CH then shares this information with the remaining CHs. A final SR is then broadcasted from the CH to its CMs.

Each cluster has to scan (C/K) (≤ 1) channels and hence the energy consumed for each cluster per sensing cycle is

$$E^C_{cent} = E^{CH}_{cent} + M_S E^S_{cent} + (M - M_S)E^{NS}_{cent} \tag{11}$$

where $M_S = \min(M, \frac{C}{K})$ is the number of scanning CMs, E^{CH}_{cent} is the energy consumed by the CH, E^S_{cent} is the energy consumed by the scanning CM, E^{NS}_{cent} is the energy consumed by the non-scanning CM.

Each CH transmits K SRs: one SR containing the scanned channel information of the cluster is unicast to each of the remaining $(K - 1)$ CHs and one SR is broadcast back to the CMs after receiving all SRs from the $(K - 1)$ CHs. Thus each CH has to decode all the M_S SRs from its CMs and $(K - 1)$ SRs from the other CHs. For simplicity of evaluation, we consider a basic access mode of IEEE 802.11 [29] without RTS/CTS and ignore the energy for the ACK in the case of unicast transmission. For simplification, we further ignore the energy for the DIFS.

$$E_{cent}^{CH} = KE_{srt} + (M_S + (K - 1))E_{srd} \tag{12}$$

Each CM transmits an SR to the CH after scanning (C/KM_S) channels and decodes the final SR that it receives from its CH.

$$E_{cent}^{S} = \left(\frac{C}{KM_S}\right) E_{scan} + \left(\frac{C}{KM_S} - 1\right) E_{sw} + E_{srt} + E_{srd} \tag{13}$$

The energy consumed by the non-scanning CMs is

$$E_{cent}^{NS} = E_{srd} \tag{14}$$

$$E_{cent} = KE_{cent}^{C} \tag{15}$$

Substituting Equations (11–14) in (15) gives the energy equation for the whole network of nodes in the centralized scheme:

$$E_{cent} = CE_{scan} + (C - KM_S)E_{sw} + K(M_S + K)E_{srt}$$
$$+ K(M_S + M + K - 1)E_{srd} \tag{16}$$

Similar to the distributed scheme, if $M > (C/K)$, then not every CM has to scan in the cluster and hence we have the following two cases:

If $M \le (C/K)$ then

$$E_{cent} = CE_{scan} + (C - KM)E_{sw} + K(M + K)E_{srt}$$
$$+ K(2M + K - 1)E_{srd} \tag{17}$$

If $M > (C/K)$ then

$$E_{cent} = CE_{scan} + (C + K^2)E_{srt} + (C + K(M + K - 1))E_{srd} \tag{18}$$

3.5 α-Centralized Cluster Based Sensing Scheme

In this scheme, only αM CMs in each cluster do the scanning while the remaining $(M - \alpha M)$ CMs do not scan for that SP.

αM can neither be less than 1 and nor can it be greater than (C/K). (C/K) is the number of assigned channels for each cluster which would be the required number of scanning nodes too. Translating this to a mathematical bound gives the following:

$$\left(\frac{1}{M}\right) \leq \alpha \leq \min\left(\frac{C}{KM}, 1\right)$$

Because of this α bound we can completely rule out the possibility of $M_S > (C/K)$ in this scheme.

Substituting $M_S = \alpha M$ in Equations (11–15) gives the following energy equation for the whole network of nodes in the α-centralized scheme:

$$E_{\alpha-cent} = CE_{scan} + (C - K\alpha M)E_{sw} + K(\alpha M + K)E_{srt}$$
$$+ K((1 + \alpha)M + (K - 1)E_{srd} \tag{19}$$

In the subsequent section, we analyse the energy savings and optimal values of α and N based on the above derived equations.

3.6 Energy Savings

Lemma 1. *Distributed cooperative sensing scheme is more energy-efficient than a non-cooperative scheme only when the number of nodes N is greater than one and the communication energy is less than*

$$\frac{(NC - C)E_{scan} + (NC - N - C + N_S)E_{sw}}{NN_S}$$

Proof. The distributed scheme saves energy over the non-cooperative scheme if and only if the total energy consumption for whole of network of nodes in the distributed scheme is less than the total energy consumption for the same network of nodes in the non-cooperative scheme. This is represented by the following relation:

$$E_{dist} \leq E_{non-coop} \tag{20}$$

Since communication is possible only when the network has more than one node, for the distributed scheme to be valid and Equation (20) to hold, the following condition should be met:

$$N > 1$$

Solving Equation (20) using Equations (1) and (5) gives the following condition for communication energy:

$$E_{scr} \leq \frac{(NC - C)E_{scan} + (NC - N - C + N_S)E_{sw}}{NN_S}$$

where $E_{src} = E_{srt} = E_{srd}$, $N_S = \min(N, C)$ for the distributed scheme, and $N_S = \min(\alpha N, C)$ for the α-distributed scheme.

For ease of simplification, E_{srt}, E_{srd} in Equation (5) are jointly denoted by E_{src} which is the communication energy in general.   □

Lemma 2. *Centralized cluster based cooperative sensing scheme is more energy-efficient than a non-cooperative scheme only when the number of clusters and the number of cluster members is at least one and the communication energy is less than*

$$\frac{(NC - C)E_{scan} + (NC - N - C + KM_S)E_{sw}}{K(2M_S + 2K + M - 1)}$$

Proof. The centralized scheme saves energy over the non-cooperative scheme if and only if the total energy consumption for whole of network of nodes in the centralized scheme is less than the total energy consumption for the same network of nodes in the non-cooperative scheme. This is represented by the following relation:

$$E_{cent} \leq E_{non-coop} \tag{21}$$

For a network to form a cluster, there should at least be one cluster and one cluster member in the cluster and hence we have

$$K \leq 1, \quad M \leq 1$$

Solving Equation (21) using Equations (1) and (16) gives the following condition for the communication energy:

$$E_{src} \leq \frac{(NC - C)E_{scan} + (NC - N - C + KM_S)E_{sw}}{K(2M_S + 2K + M - 1)}$$

where $E_{src} = E_{srt} = E_{srd}$, $M_S = \min(M, C/K)$ for the centralized scheme, and $M_S = \min(\alpha, M, C/K)$ for the α-centralized scheme.   □

3.7 Optimal α Values for the α-Distributed Scheme

3.7.1 Optimal α for Maximum Energy Savings

The optimal α for a given N is defined as the value of α where the energy savings of the α-distributed scheme over the distributed are the maximum. The energy savings can be looked at, from two different perspectives – the energy savings for the whole network of nodes and the energy savings per scanning node in the network. We first look at the optimal α for maximum energy savings considering the whole network of nodes over one sensing cycle paired with various constraints to form a set of optimization problems.

Optimization Problem 1: Optimal α considering the energy savings for the entire network of nodes:

$$\text{Maximize} f(\alpha) = \left(\frac{E_{dist} - E_{\alpha-dist}}{E_{dist}} \right) \text{ such that } \left(\frac{1}{N} \right) \le \alpha \le \min \left(\frac{C}{N}, 1 \right)$$

The α value we consider optimal here, is the value of α where the α-distributed scheme shows the maximum energy savings over the distributed scheme considering the entire network of nodes for a given sensing cycle. $f(\alpha)$ is maximized at $\alpha = (1/N)$ which thus becomes the optimal α. However, considering the shadowing phenomenon, battery characteristics of the wireless nodes and most importantly the limited scanning and reporting periods, this value should be chosen with discretion in order to avoid poor spectral efficiencies and accuracies, shorter operating lifetimes respectively [28]. Hence a constraint to limit the sensing and reporting time in a given sensing cycle is needed as shown below in the next optimization problem.

Optimization Problem 2: Optimal α considering the energy savings for the entire network of nodes along with the time constraint:

$$\text{Maximize } f(\alpha) = \left(\frac{E_{dist} - E_{\alpha-dist}}{E_{dist}} \right)$$

$$\text{such that } \left(\frac{1}{N} \right) \le \alpha \min \left(\frac{C}{N}, 1 \right) \text{ and } L \le 1$$

where L is the total time for the SC as defined below:

$$L = \left(\frac{C}{\alpha N} \right) T_{scan} + \left(\frac{C}{\alpha N} - 1 \right) T_{sw} + \alpha N T_{data} \tag{22}$$

In the evaluation section, we plot the optimal values where $f(\alpha)$ is maximized and the constraint $L \le l$ ms is satisfied.

Also, for further study of the relative energy cost comparison of a scanning node in distributed scheme versus a scanning node in α-distributed scheme, we maximize the function $f(\alpha)$ with a scanning node energy constraint.

Optimization Problem 3: Optimal α considering the energy savings for the entire network of nodes along with the per scanning node energy constraint:

$$\text{Maximize } f(\alpha) = \left(\frac{E_{dist} - E_{\alpha-dist}}{E_{dist}} \right)$$

$$\text{such that } \left(\frac{1}{N} \right) \leq \alpha \leq \min\left(\frac{C}{N}, 1 \right) \text{ and } g(\alpha) \geq 0$$

where

$$g(\alpha) \geq 0 = \left(\frac{E_{dist}^S - E_{\alpha-dist}^S}{E_{dist}^S} \right)$$

The $g(\alpha) \geq 0$ constraint was taken into account considering the fact that for a given sensing cycle, although the α-distributed scheme saves energy for the network of nodes N on the whole, it should not burden each scanning node with a very high number of channels to be scanned, causing the node to die quicker. This constraint makes sure that the energy of the scanning nodes in the α-distributed scheme is either less than or at least equal to the energy of scanning node in the distributed scheme. The optimal α with this constraint is always the upper bound of α given by $\min(C/N, 1)$. However if the constraint is relaxed to be less than zero, i.e., $g(\alpha) \leq 0$, the optimal α would move to the lower bound which is $(1/N)$. $g(\alpha)$ thus helps to explain the fact that the per scanning node energy cost of α-distributed scheme versus that of the distributed over one sensing cycle is always higher except at $\min(C/N, 1)$ where it equals the distributed. In the next optimization problem, we show that the energy savings over a cumulative sensing cycle period are however positive.

Optimization Problem 4: Optimal α considering energy savings for the entire network of nodes along with the per scanning node energy constraint over n sensing cycle period:

$$\text{Maximize } f(\alpha) = \left(\frac{E_{dist} - E_{\alpha-dist}}{E_{dist}} \right)$$

$$\text{such that } \left(\frac{1}{N} \right) \leq \alpha \leq \min\left(\frac{C}{N}, 1 \right) \text{ and } h(\alpha) \geq 0$$

where

$$h(\alpha) = \left(\frac{\min\left(\frac{C}{N}, 1 \right) E_{dist}^S - (\alpha E_{\alpha-dist}^S)}{\min\left(\frac{C}{N}, 1 \right) E_{dist}^S} \right)$$

The constraint $h(\alpha)$ defines that, for a given n sensing cycle period each scanning node in the distributed scheme would have to scan $\min(C/N, 1) \times n$ times on an average while the scanning node in the α-distributed scheme would have to scan only for $\alpha \times n$ times on an average.

With this new constraint, the optimal α value is still $(1/N)$. However, the numerical evaluation results of $h(\alpha)$ show that the constraint itself results in positive values unlike the constraint $g(\alpha)$. This goes on to show that the α-distributed scheme saves energy from a scanning node's perspective as well as for the whole network over a given period of n sensing cycles and more noticeably at smaller α values.

3.7.2 Optimal N for Maximum Energy Savings

It is also interesting enough to analyse if there is an optimal N for a given α where the energy savings of the α-distributed over distributed are maximum. The next two optimization problems discuss the same.

Optimization Problem 5: Optimal N considering energy savings for the entire network of nodes over one sensing cycle:

$$\text{Maximize } F(N) = \left(\frac{E_{dist} - E_{\alpha-dist}}{E_{dist}} \right) \text{ such that } \left(\frac{1}{\alpha} \right) \le N \le \min \left(\frac{C}{\alpha}, 1 \right)$$

Solving $F(N)$ for maximization, shows that it is maximized at

$$N = \max \left(\frac{-E_{scan} + \sqrt{E_{scan}^2 + E_{sw}E_{scan} + \left(\frac{C}{\alpha} \right) E_{src}E_{scan} + \left(\frac{C}{\alpha} \right) E_{sw}E_{src}}}{E_{src}}, C \right)$$

Similarly the next optimization problem looks at maximizing the per node energy savings over n sensing cycles.

Optimization Problem 6: Optimal N considering the energy savings per node over n sensing cycles:

$$\text{Maximize } G(N) = \left(\frac{\min \left(\frac{C}{N}, 1 \right) E_{dist}^S - (\alpha E_{\alpha-dist}^S)}{\min \left(\frac{C}{N}, 1 \right) E_{dist}^S} \right)$$

$$\text{such that } \left(\frac{1}{\alpha} \right) \le N \le \min \left(\frac{C}{\alpha}, 1 \right)$$

Using the Lingo optimization package, evaluating the function $G(N)$ for the highest magnitude of energy savings per scanning node over n sensing cycles shows that it is maximized at $N = C$.

The values showing the trend of optimal N over a varying range of α for functions $F(N)$ and $G(N)$ are plotted separately in the next section.

The outcomes of all the above optimizations are completely dependent on α, which is the basis of the relation between distributed and α-distributed. Since this relation remains the same between centralized and α-centralized, the optimization results and hence the inferences are going to be similar. So we do not look at optimizations for the centralized schemes in this work.

4 Evaluation

We do a numerical evaluation for all the proposed generic equations and show the energy savings of the α-schemes over their counterparts. All the base values for E_{scan}, E_{sw}, E_{srt}, E_{srd} are calculated from equations and values specified in Table 1 [9]. Following [25] where White-Fi has to scan at least about 50 TV channels and [26] where the notion of channels can just be sub bands obtained by dividing a given wide band, we believe $C = 100$ would be a suitable and practical value for the number of channels to be scanned. In [27] the authors show that the optimal sensing time for a SU to detect the PU with 90% probability is about 15 ms and in [28] the false alarm probability had a linear down trend as scan time was varied from 20 to 100 ms. So we infer that $T_{scan} = 50$ ms would be an appropriate value to achieve a good detection probability and low false alarm rates simultaneously. We also make the assumption that scanning takes longer time than data transmission, which holds true for most schemes practically except perhaps clear channel assessment (CCA) based schemes. We evaluate the total energy consumed for each of the schemes by varying the range of number of nodes N and the fraction factor α.

4.1 Distributed versus α-Distributed Schemes

Figure 7 shows the energy trends over a varying N for the distributed and α-distributed schemes. The lower the value of α, the lower are the energy costs for the α-distributed scheme. These energy savings become more apparent for higher node densities. However at the point the energy cost of the α-distributed scheme equals the distributed scheme and hence the energy savings decrease from positive values to zero. This is due to the fact that

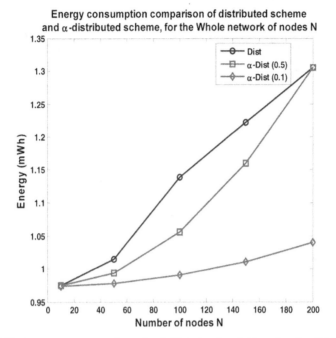

Figure 7 Total energy consumption of whole network of N nodes in distributed and α-distributed schemes.

the significance of the fraction factor α lies only in the range suggested in Equation (6), further explaining the point that there is no reason to have scanning nodes greater than the required number, which would be the number of channels C for both the distributed and the α-distributed schemes. Hence our proposed α-distributed scheme holds significance as long as $\alpha N < C$.

Figure 8 shows that the average energy costs of a node for both distributed and α-distributed keep decreasing with increasing node densities. A closer analysis indicates that this is due to the reduction in the scanning energies; since with increasing node densities, the number of channels scanned on an average by each node keeps decreasing. Though the reporting energy increases on an average, this increase is greatly offset by the decrease in the scanning energy since $E_{scan} \ll E_{srt}$ (or) E_{srd}.

4.2 Optimal Values

The optimal α values for a given N where $f(\alpha)$ is maximized and the constraint of the sensing cycle time length L is satisfied are shown in Figure 9.

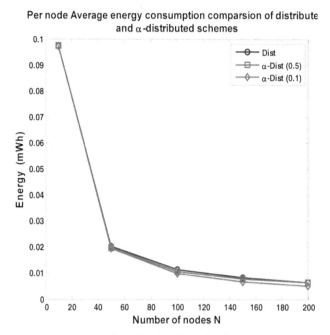

Figure 8   Average energy consumption of each node in distributed and α-distributed schemes.

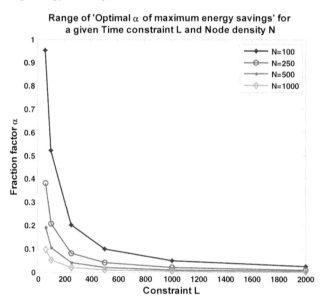

Figure 9   Optimal values of α where the energy savings of the α-distributed scheme over the distributed scheme are maximum, for a given N and L.

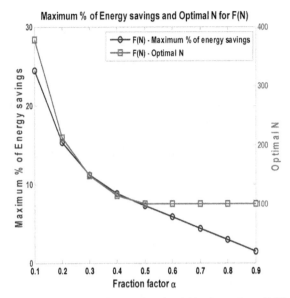

Figure 10 Percentage of energy savings and optimal N values where $F(N)$ is maximized.

The optimal values were computed in the Lingo optimization package using the formulations developed. It can be clearly noticed that the optimal α values decrease with increasing L. With higher L each node gets to scan more channels and so lesser scanning nodes are needed which results in a smaller α.

A further study of optimal values of N for a given α can be done using Figures 10 and 11. The 'maximum % of energy savings' line for both plots is for the corresponding energy savings at that optimal N for $F(N)$ and $G(N)$. The function $F(N)$ has the highest magnitude of energy savings at various values of N for varying α until $\alpha = 0.5$, after which the optimal N stays at 100 (value of C). Optimal N for function $G(N)$ is always at $N = C = 100$ regardless of the value of α. The energy savings for both $F(N)$ and $G(N)$ predictably go down with increasing α values.

4.3 Centralized versus α-Centralized Schemes

Figure 12 shows the energy trends over a varying N for the centralized and α-centralized schemes. As expected, the centralized schemes in general have lower energy costs than the distributed schemes and the α-centralized schemes have lower energy costs over the centralized scheme.

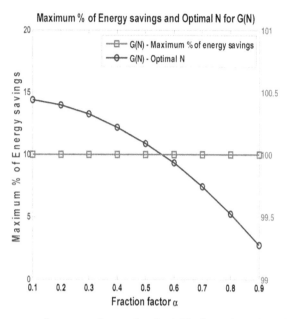

Figure 11   Percentage of energy savings and optimal N values where $G(N)$ is maximized.

The α-centralized schemes have lower energy costs with decreasing α and this can attributed to the lesser control overheads both for the CH and the CMs. Also, the energy increase with increasing N is more linear in the centralized schemes while this increase is inclined towards being exponential in the distributed schemes. This clearly shows that centralized schemes in general are more energy efficient and hence should be the first choice at higher N values. The values used in this plot were derived for $K = 1$. To gain further insight into the impact of K on energy consumption, we use Figure 13 which shows that a higher K results in higher energy overhead.

4.4 Non-Cooperative versus Cooperative Schemes

Cooperative schemes not only reduce the bandwidth requirements to convey the scanning information but also the energy consumed to scan and share this information. Figure 14 proves this claim and it can be noticed that a logarithmic scale was used to capture the wide variation of the energy values of non-cooperative schemes and the lower energy values of the cooperative schemes. The centralized schemes were plotted for $K = 1$ for a fair comparison.

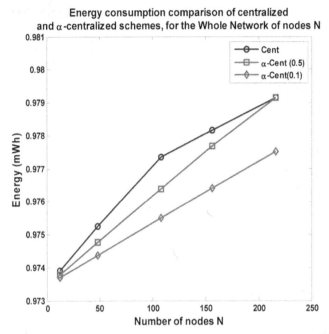

Figure 12 Total energy consumption for whole network of N nodes in centralized and α-centralized schemes.

5 Discussion and Conclusions

The energy model of sensing developed in this work gives a platform for energy accountability, quantification and comparison of the non-cooperative sensing schemes and the generic cooperative sensing schemes – distributed and centralized, along with our proposed new α-schemes. Our investigation on their energy costs shows that the cooperative schemes outperform the non-cooperative scheme under the simplifying assumptions made and further the α-schemes are significantly energy efficient than the generic schemes. Optimal values for the fraction factor α and number of nodes N derived contribute to further useful insights on the relative energy savings.

Our results make some idealized and simplifying assumptions to make the analysis tractable in taking the first step of comparing cooperative and non-cooperative schemes for CRs. The current network scenario is considered to be a clique; a non-clique network would require additional work for cooperation relying on multi-hop communication which will increase the energy cost for cooperative schemes. We further make the assumption of perfect

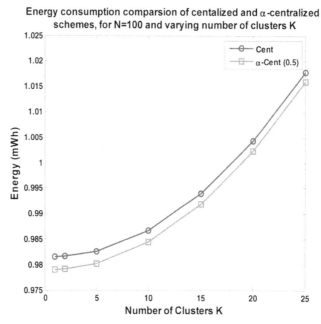

Figure 13 Impact of K on the total energy consumption for whole network of N nodes in centralized and α-centralized schemes.

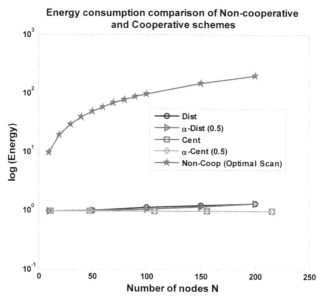

Figure 14 Energy consumption comparison of the non-cooperative and cooperative schemes.

scheduling in the sensing model; this assumption is feasible in networks that are tightly time-synchronized which is relatively easy to achieve in static networks with nodes within range of each other. The impact of achieving synchronization will have to be considered for networks that have mobile nodes, or require multi-hop communications. Finally, we do not consider power management issues in this paper. Our results are complementary to energy savings possible by such schemes. The results presented in this paper are based only on the energy to scan and energy to communicate. A useful next step will be to tie in the impact of various power management schemes and deal with synchronizing communication schedules and factor in energy costs.

References

[1] Cognitive Radio Research-Project overview, Berkeley Wireless Research Center (BWRC), http://bwrc.eecs.berkeley.edu/Research/Cognitive/home.htm.
[2] Spectrum occupancy measurements performed by Shared Spectrum Company, http://www.sharedspectrum.com/papers/spectrum-reports, Chicage, Illinois, 2005.
[3] J. Mitola and G.Q. Maguire. Cognitive radio: Making software radios more personal. *IEEE Personal Commun. Mag.*, 6:13–18, 1999.
[4] D. Cabric, A. Tkachenko, and R. Brodersen. Spectrum sensing measurements of pilot, energy, and collaborative detection. In *Proceedings of IEEE Military Communications Conference*, October 2006.
[5] D. Cabric, S. Mishra, and R. Brodersen. Implementation issues in spectrum sensing for cognitive radios. In *Proceedings of Asilomar Conference on Signals, Systems and Computers*, Vol. 1, 2004.
[6] R. Tandra and A. Sahai. Fundamental limits on detection in low SNR under noise uncertainty. In *Proceedings of IEEE International Conference Wireless Networks, Communications and Mobile Computing*, 2005.
[7] M. Oner and F. Jondral. Cyclostationarity based air interface recognition for software radio systems. In *Proceedings of IEEE Radio and Wireless Conference*, September 2004.
[8] A. Ghasemi and E.S. Sousa. Asymptotic performance of collaborative spectrum sensing under correlated log-normal shadowing. *IEEE Commun. Lett.*, 2007.
[9] V. Namboodiri. Are cognitive radios energy efficient? A study of the wireless LAN scenario. In *Proceedings of IEEE International Computing and Communications Conference*, 2009.
[10] Wenzhong Wang and Luyong Zhang. On the distributed cooperative spectrum sensing for cognitive radio. In *Proceedings of International Symposium on Communication and Information Technologies*, 2007.
[11] N. Ahmed, D. Hadaller, and S. Keshav. GUESS: Gossiping updates for efficient spectrum sensing. In *Proceedings of International Workshop on Decentralized Resource Sharing in Mobile Computing and Networking*, 2006.

[12] J. Nicholas Laneman. Cooperative diversity in wireless networks: Efficient protocols and outage behavior. *IEEE Transactions on Information Theory*, 50(12), December 2004.

[13] Q. Chen and M. Motani. Cooperative spectrum sensing strategies for cognitive radio mesh networks. *IEEE Journal of Selected Topics in Signal Processing*, 2011.

[14] Chia-han Lee. Energy efficient techniques for cooperative spectrum sensing in cognitive radios. In *Proceedings of IEEE Conference on Communications and Networking*, 2008.

[15] Chunhua Sun, Wei Zhang, and K. Ben. Cluster-based cooperative spectrum sensing in cognitive radio systems. In *Proceedings of IEEE International Conference on Communications*, 2007.

[16] Chen Guo and Tao Peng. Cooperative spectrum sensing with cluster-based architecture in cognitive radio networks. In *Proceedings of Vehicular Technology Conference*, 2009.

[17] Bin Shen and Chengshi Zhao. User clusters based hierarchical cooperative spectrum sensing in cognitive radio networks. In *Proceedings of IEEE International Conference on Cognitive Radio Oriented Wireless Networks and Communications*, 2009.

[18] Youmin Kim and Wonsop Kim. Group-based management for cooperative spectrum sensing in cognitive radio networks. In *Proceedings of International Conference on Advanced Communication Technology*, 2010.

[19] R. Akhtar, A. Rashdi, and A. Ghafoor. Grouping technique for cooperative spectrum sensing in cognitive radios. In *Proceedings of International Workshop on Cognitive Radio and Advanced Spectrum Management*, 2009.

[20] Chunmei Qi and Jun Wang. Weighted-clustering cooperative spectrum sensing in cognitive radio context. In *Proceedings of International Conference on Communications and Mobile Computing*, 2009.

[21] Duan, Jiaqi and Li, Yong. A novel cooperative spectrum sensing scheme based on clustering and softened hard combination. In *Proceedings of International Conference on Wireless Communications, Networking and Information Security*, 2010.

[22] The difference engine: Bigger than Wi-Fi. Available from http://www.economist.com/blogs/babbage/2010/09/white-space_wireless, accessed on 23 September 2010.

[23] Chloe Albanesius. FCC opens TV 'white spaces' for unlicensed super Wi-Fi. Available from http://www.pcmag.com/article2/0,2817,2369580,00.asp, accessed on 23 September 2010.

[24] S. Xie, Y. Liu, Y. Zhang, and R. Yu. A parallel cooperative spectrum sensing in cognitive radio networks. *IEEE Transactions on Vehicular Technology*, 2010.

[25] L. Verma, Sim Daeyong, and S.S. Lee. Wireless networking in TV white space leveraging Wi-Fi. In *Proceedings of IEEE 14th International Symposium on Consumer Electronics*, 2010.

[26] A.R. Biswas, T.C. Aysal, S. Kandeepan, D. Kliazovich, and R. Piesiewicz. Cooperative shared spectrum sensing for dynamic cognitive radio networks. In *Proceedings of IEEE International Conference*, 2009.

[27] Ying-Chang Liang, Yonghong Zeng, E.C.Y. Peh, and Anh Tuan Hoang. Sensing-throughput tradeoff for cognitive radio networks. *IEEE Transactions on Wireless Communications*, 2008.

[28] Hyoil Kim and K.G. Shin. Efficient discovery of spectrum opportunities with MAC-layer sensing in cognitive radio networks. *IEEE Transactions on Mobile Computing*, 2008.

[29] IEEE 802.11 wireless local area networks standard-IEEE Standard for Information technology – Telecommunications and information exchange between systems – Local and metropolitan area networks. Available from http://standards.ieee.org/getieee802/download/802.11-2007.pdf.

Biographies

Reshma Syeda is a recent graduate of Wichita State University. She completed her M.S. degree in Electrical Engineering in 2011 under the supervision of Professor Vinod Namboodiri. She has also worked as a graduate research assistant at the Cisco Technical Research Center of WSU. Reshma received her B.Tech degree in Electronics and Communications Engineering from Jawaharlal Nehru Technological University, India in 2007 and has also worked for Infosys Technologies Ltd. for two years as a software developer. Her research interests include cognitive radio networks and smart grids. She is currently working as a software engineer at NetApp.

Vinod Namboodiri is currently an Assistant Professor at the Department of Electrical Engineering and Computer Science at Wichita State University, USA. He graduated with a Ph.D. in Electrical and Computer Engineering from the University of Massachusetts Amherst in 2008. He has served or is currently serving on the technical program committees of IEEE GLOBE-COM, IEEE ICC, IEEE IPCCC, and IEEE GREENCOM, and is an active reviewer for numerous journals and conferences in the mobile computing and green communications areas. His research interests include designing algorithms and protocols for energy-intelligent and sustainable computing, and designing effective communication architecture for smart electric grids. In the past he has worked on designed MAC and routing protocols, and developing energy-efficient protocols and algorithms for different wireless technologies like Wireless LANs, RFID Systems, Wireless Sensor Networks, and Wireless Mesh Networks.

Energy Efficiency and Capacity Modeling for Cooperative Cognitive Networks

Rasool Sadeghi, João Paulo Barraca and Rui L. Aguiar

Instituto de Telecomunicações, Universidade de Aveiro, 3810-193 Aveiro, Portugal;
e-mail: rsadeghi@av.it.pt, {jpbarraca, ruilaa}@ua.pt

Received 7 October 2011; Accepted: 19 April 2012

Abstract

Cooperative relaying has recently appeared as one of the widely recognized features for future wireless communication systems. The great potential of cooperative communication in increasing system capacity and enhancing power efficiency has attracted large efforts over the last few years. In this paper, we propose a Cooperation Loop as a reference model for all algorithms in relay based cooperative wireless networks. Using this model, we discuss cooperative relay based protocols in IEEE 802.11 standards and limits posed to cognitive approaches. We show the potential location area of relay nodes as well as the performance bounds of capacity gain, delay and power efficiency achieved in relay based scenarios for any cooperative cognitive algorithms.

Keywords: cooperation, cognitive, IEEE 802.11 MAC protocols, capacity, energy efficiency.

1 Introduction

Cooperative communications based on relay nodes have recently emerged as a novel approach in the design of next-generation wireless networks [1–8]. The classical paradigms of point-to-point and point-to-multipoint in wireless networks are being replaced by new interactions models, where the nodes cooperate with each other in order to improve the performance of their

Journal of Green Engineering, Vol. 2, 377–399.

own communication and of the global network. This aspect is of particular importance if done to maximize delivery under a varying environment, and employing cognitive techniques. The multipath propagation feature of the radio communication medium, long considered as the main reason of interference in conventional wireless networks, is now regarded as a potential resource for possible performance improvement in cooperative relay based networks, as well a source of potential energy reductions. In this concept, neighboring nodes overhear other messages and potentially help by relaying information. Cooperative relay communications address main challenges [2, 3, 9–12] in different types of wireless networks with the purpose of improving a given metric, such as overall system performance, or energy efficiency, by increasing capacity, survivability, range, and throughput, or simply transmission efficiency. One important aspect of cooperation is that cooperation is not always beneficial. Cognitive mechanisms must be used to evaluate the current environment and decide whether cooperation brings any improvement to network operation.

Recently, the topic of cognitive and cooperative networking has received significant attention from researchers, in particular when considering the IEEE 802.11 standard [13]. The IEEE 802.11 family of protocols arose as the dominant industrial standard for Wireless Local Area Networks (WLANs) providing simple mechanisms for the establishment of either infrastructure or ad-hoc networks. By allying cooperative and efficient cognitive schemes it is possible to devise promising solutions to improve the main features of the IEEE 802.11 standard, such as multiple transmission data rate adaptation and power control mechanisms. This highlights the potential practical role of cooperation to save the common network resources of power and spectrum. This work addresses this subject and has two major purposes: (1) To investigate the performance bounds of capacity gain obtained by cooperation in IEEE 802.11 networks, which can be used by a cognitive algorithm to decide when and how to cooperate, and (2) to assess power efficiency and power gain bounds of cooperative schemes in the best case scenario for cooperation, in such a way that can be explored in practical cognitive algorithms.

The rest of the paper is organized as follows: The concept of cooperation loop and the associated relevant parameters in IEEE 802.11 are described in Section 2. Section 3 discusses how cooperation can provide the solution for the main challenges in IEEE 802.11 MAC. Section 4 considers the mathematical methods to calculate the performance bounds of delay and capacity in IEEE cooperative relay based networks. In Section 5, we present a mathematical analysis evaluating the power gain and energy efficiency of relay based

MAC protocols, that can be incorporated in cooperative cognitive algorithms. Section 6 presents our simulation results, while Section 7 concludes the paper and presents future directions.

2 Solutions for Cooperative Communications

One of the main features of IEEE 802.11 WLAN is the support for multiple transmission data rates, which are related to the instantaneous conditions of the wireless channel, terminal capabilities, performance requirements, spectrum requirements, energy constrains, or simply administrative policies. Even though this feature increases the coverage area of wireless communication, it decreases the energy efficiency of the network, and leads to the problem called performance anomaly [14]: equal transmission opportunity provided to all involved nodes in the same IEEE 802.11 network leads to high latency required to complete the transmission of low data rate nodes, thus degrading the performance of the remaining, higher rate nodes. As an example, the duration time for the transmission of a packet of fixed size at the minimum data rate (6 Mbit/s), using the IEEE 802.11g protocol [15], makes the shared medium being occupied nine times longer when compared to the transmission the same packet at highest data rate (54 Mbit/s). This problem can be exacerbated in the most recent amendment of the standard, such as IEEE 802.11n [16], which supports up to 300 Mbit/s. Therefore the overall system performance is constrained by the ratio of low data rate nodes to all nodes in the same collision domain. Furthermore, the nodes at the edge of coverage area suffer from high packet loss rate due to worse channel conditions and higher interference levels. Cooperative protocols provide promising solutions to overcome these challenges of IEEE 802.11 networks. The key idea is that devices can sense their environment, and decide to replace one channel with bad conditions by two good channels. The meaning of bad and good channels depends on the purpose of the cooperation. As an example, in CoopMAC [17] and rDCF [18] one low data rate direct transmission link can be replaced by two faster transmission links, employing a relay node, yielding higher capacity. In other words, if the throughput improvement is the main of cooperation, the good channels are the ones that reduce the transmission delay. Interested readers can find proposals addressing this topic in [19–23].

The energy efficiency of the IEEE 802.11 MAC protocol is another important aspect of communications, especially for ad-hoc nodes powered by batteries and or other sources, such as solar panels. Therefore, efficient utilization of energy is a main concern of MAC protocol designers and the awareness to g green is now widely popular. Even though most of research works deal with throughput improvement gained by relay based MAC protocols, a few publications [17, 24–27] focus on the impact of relay nodes on the energy efficiency in cooperative relay based IEEE 802.11 networks. The authors of [24] demonstrated that cooperative relay schemes provides the power saving ranging from 7 to 20 dB over direct transmission and from 1 to 3 dB over multihop routes. It was observed in [17] that besides the network capacity gain by cooperative communications, energy efficiency gain is on the order of 20–40%. The energy efficiency of MIMO and cooperative MIMO systems were also investigated in [25] in which the authors address the issues such as energy cost and reduction of transmitter power in cooperative relay schemes. In addition, the authors in [26] demonstrate the energy efficiency of single relay cooperative MAC protocol while the results in [27] indicate the energy saving mechanism and energy performance improvement of multiple relay protocol when compared to normal IEEE 802.11 MAC protocol. Nevertheless, none of these works is able to provide a practical view of when to use cooperation, in the sense of a framework and rules guidelines that can be incorporated in the cognitive algorithms running on each node and that consider rate adaptation and energy efficiency simultaneously.

3 Cooperation in IEEE 802.11

Cognitive systems are able to adjust their operation according to changes in their environment. Therefore, similar to Autonomic Systems [28], cognitive systems make use of information sensed, which serves as feedback for future decisions. Our proposed Cooperation Loop can be structured as depicted in Figure 1, consisting of three phases: sense, decide and act. In this architectural reference model, the inception of cooperation is carried out, at each node, by sensing methods to sense the environment and neighbor nodes (Sense). The observation captured by the sense phase will be further used for a cognitive decision (Decide) when the cooperative strategies are determined based on cooperation policies, available sensing parameters and cooperation purposes. The final phase fulfills the cooperation procedure by sending the required control messages and initiating the cooperative transmission (Act). Different wireless communication environments such as Wireless LANs (WLANs), Wireless Sensor Networks (WSNs), Wireless Mesh Networks (WMNs), and Wireless Cellular Networks (WCNs) have particular requirements, which can be mapped to the operation of all states in this reference model. This cooper-

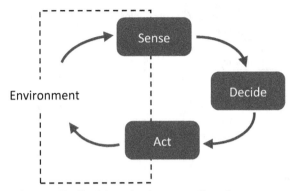

Figure 1 Feedback loop for the cooperative relay process.

ation loop concept can be applied for spectrum of features from channel level to network level and negotiating procedures within and across the OSI layers.

Since our study is focused on cooperation in the IEEE 802.11 family of standards, Table 1 summarizes different options for different cooperation loop phases. Some sensing parameters are achieved by overhearing the ongoing transmissions and extract the intended information explicitly from MAC header frames (e.g. RSSI [18]) or can be obtained by some computation based on overheard packets (e.g. data rate [12]). Moreover, sensing of some events such as packet failure and collision occurrence, can be another method to overhear the environment and initiate the cooperation if there are some possibilities. However, there are some protocols to exploit the cooperation by using upper layer features and cross layer approach: in [34], the priority of traffic flows is mapped in IEEE 802.11 MAC control packets and the overhearing nodes can sense the priority of ongoing packets and compare to their own priorities [34]. The Decision phase of cooperative MAC protocols includes parameters such as the number of employed relay(s) and a set of relay selection metrics. The Act phase determines which node(s) can initiate the cooperation and how the control mechanism and notification signaling are applied. This spectrum of parameters can be used by a myriad of cognitive algorithms for cooperative communication, combined differently depending on the overall system objectives. In the scope of cognitive networks, sensed information and past decisions can also be used to make future decisions, or to other control layers (The set of sensed parameters can include information from applications. In particular, what delivery requirements they have, or what type of power source is available.)

Table 1 Summary of cooperation loop in IEEE 802.11.

Sensing	Decision		Act	
Type	Relay selection metric	Initiation	Control	Notification
RSSI [19]	Max. data rate [18]	Source [19]	Centralized [18]	CFC
PLCP header [18]	Random [32]	Relay [36]	Distributed [19]	Message [32]
Packet Failure [32]	Priority based [35]	Source-	Hybrid [36]	
Data rate [30]	Service	Relay [35]		
Collision occurrence [12]	differentiation [34]			
Priority of traffic flow [34]				

Unlike simple cognitive radio communications [29] which are based on spectrum sensing, and reconfigurable capabilities, we propose the usage of cooperation loop as a modular framework, able to support different cooperative cognitive protocols for existing IEEE 802.11 networks, according to the best interest of a particular network or individual nodes. Depending on the main purpose of cooperation, we can select one or some of the sensing parameters listed in Table 1 and then we can create a composite metric. By using this metric the decider node determines which set(s) of nodes can participate in cooperation and also estimates which set of source, destination and relay(s) can be more beneficial for a particular traffic. In the Act phase of cooperation loop there are spectrum of features including the initiation, signaling and control plan for a cooperative cognitive process. In some of these phases, we can also exploit opportunistic schemes instead of deterministic ones [30].

4 Delay Performance and Capacity Gain in Cooperative IEEE 802.11

In order to improve the throughput in cooperative IEEE 802.11 network using relay nodes, the transmission delayshould often considered as the main metric to initiate the cooperation process. Transmission delay is the time a data packet takes to be transmitted over the medium. So we should have a practical sensing method to obtain the transmission delay of direct and relay paths. For instance, for given three nodes as depicted in Figure 2, node R as a potential relay can explicitly obtain the actual data rate achieved between Access Point(AP) and $N(k)$ from overhearing data frames exchanged between node (AP) and N. The IEEE 802.11 MAC header (or in more detail, the PLCP sub-header), contains a field named SIGNAL, which denotes the bit rate of

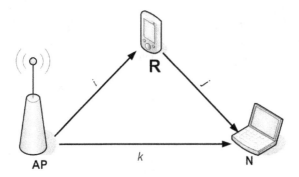

Figure 2   Cooperative scenario using relay node in infrastructure WLAN.

every data packet sent to the network. Node R can also discover the potential bitrates between itself and $AP(i)$ and $N(j)$. For that, node R measures the Received Signal Strength Indication (RSSI) of RTS and CTS frames issued by the AP and N and computes the corresponding data rates of obtained RSSI(s). After obtaining the three data rates between these three nodes (Sense phase), we can define a metric such as Delay Ratio (DR) [33]. Delay ratio is the ratio between the transmission delay of relay path, and that of the direct path (Decide phase) and can be expressed as

$$DR_{ijk} = \frac{i^{-1} + j^{-1}}{k^{-1}} \qquad (1)$$

when the relay node supports data rates of i Mbit/s and j Mbit/s to AP and node N respectively and the direct transmission data rate between AP and node N is k Mbit/s. In this paper, we focus on the bounds of MAC layer performance with high data packet size and we ignore the MAC overhead. It is noted that the accurate value of delay ratio should include the overhead and it depends on the specific cooperation technique. Clearly, if the value of the calculated delay ratio is less than 1, the relay channel will possibly provide better transmission characteristics than the direct channel, due to the resulting higher bandwidth and lower transmission delay for end-to-end communication. Note that in this first approximation, the processing delay in node R and access delay are neglected. In addition, in order to consider the capacity improvement corresponding to the obtained delay ratio, we define the Cooperative Capacity (CC) as

$$CC_{ijk} = k \frac{i^{-1} + j^{-1}}{k^{-1}} \qquad (2)$$

Table 2 Delay ratio less than 1 and equivalent cooperative capacity for IEEE 802.11b.

DELAY RATIO	AP to N Data rate (Mbit/s)	AP to R Data rate (Mbit/s)	R to N Data rate (Mbit/s)	Cooperative capacity (Mbit/s)
0.18	1	11	11	5.5
0.27	1	11	5.5	3.7
0.36	1	5.5	5.5	2.8
0.59	1	2	11	1.7
0.68	1	2	5.5	1.5
0.36	2	11	11	5.6
0.54	2	11	5.5	3.7
0.72	2	5.5	5.5	2.8
1	5.5	11	11	5.5
2	11	11	11	5.5

Table 2 shows all delay ratio values less than 1 and equivalent cooperative capacity obtained by cooperative relay based communication in scenario of Figure 2 using IEEE 802.11b (the table is limited to IEEE 802.11b for simplification). The best performance for cooperation occurs when delay ratio is minimal, in this case 0.18, and provides a reduction in delay of around 72%, considering that the direct data rate is 1 Mbit/s, and the data rate between source to Figure 2 Cooperative scenario using relay node in infrastructure WLAN relay and relay to destination is 11 Mbit/s. As it can be observed, the usefulness of using a relay decreases, when the data rate between AP and N increases. Table 2 shows that relay selection algorithms are only useful when end-to-end data rate is near the lower limits allowed by the standard. For instance, in the IEEE 802.11b standard if the data rate between AP and N is 5.5 Mbit/s or 11 Mbit/s, no cooperation leading to a reduction in delay will be possible. This is a practical rule that any cognitive algorithm for IEEE 802.11b should incorporate.

One of the main questions in cooperative relay based wireless networks is which percentage of the access point coverage area can potentially improve the performance corresponding to obtained delay ratio. Another important point of concern is what are the performance bounds (minimum, maximum) and expected average of delay ratio and capacity gain achieved by cooperative schemes in the different data rates supported by IEEE 802.11 family of protocols. In the rest of this section, we will answer these questions by proposing a mathematical model for delay and area analysis.

The relay nodes can perceive the various values of delay ratio due to different data rates supported in IEEE IEEE 802.11. The area wherein every

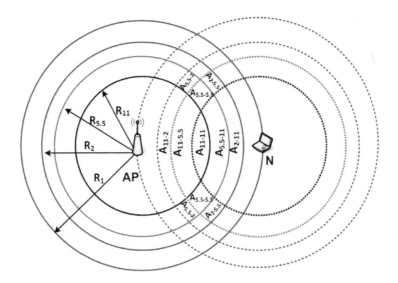

Figure 3 Relay area for direct transmission of 1 Mbit/s.

relay node can move while its delay ratio does not change is called relay area. To obtain the relay area versus delay ratio, we consider the geometric model of cooperation. As depicted in Figure 3, an infrastructure WLAN includes node N which is located at the transmission range of R_1 to support 1 Mbit/s data rate to AP. The intersection area of A_{ijk} denotes the potential area for relay node corresponding to DR_{ijk} discussed in (1). To obtain the $A_{ijk}'s$, we consider the overlap area of two circles with radii of r_1 and r_2 and distance of l between their centers. The overlap area, denoted by S_{r1r2} can be written as

$$S_{r1r2} = r_1^2 \sin^{-1}(h/r_1) + r_2^2 \sin^{-1}(h/r_2) - hl \qquad (3)$$

where

$$h = \frac{\sqrt{2r_1^2 r_2^2 + 2(r_1^2 + r_2^2)l^2 - (r_1^4 + r_2^4) - l^4}}{2l}$$

The relay area is not covered directly by calculated overlap area, but the relation between this overlap area of two circles and relay area of $A_{ijk}'s$ in Figure 3 can be easily calculated as Equation (4). The value of k is equal to 1 because of direct transmission data rate of 1 Mbit/s between AP and N in Figure 3. Table 3 summarizes all relay area and corresponding delay ratio

Table 3 Delay ratio of different relay area for direct transmission of 1 Mbit/s.

Relay area	$A_{11-11-1}$	$A_{11-5.5-1}$	$A_{5.5-5.5-1}$	A_{11-2-1}	$A_{5.5-2-1}$
Delay ratio	0.18	0.27	0.36	0.59	0.68

values.

$$\begin{cases} A_{11-11-1} = S_{R_{11}R_{11}} \\ A_{11-5.5-1} = A_{5.5-11-1} = S_{R_{11}R_{5.5}} - S_{R_{11}R_{11}} \\ A_{11-2-1} = A_{2-11-1} = S_{R_{11}R_2} - S_{R_{11}R_{5.5}} \\ A_{5.5-5.5-1} = (S_{R_{5.5}R_{5.5}} - S_{R_{11}R_{5.5}} - A_{11-5.5-1})/2 \\ A_{5.5-2-1} = A_{2-5.5-1} = (S_{R_{5.5}R_2} - S_{R_{5.5}R_{5.5}} - A_{2-11-1})/2 \end{cases} \qquad (4)$$

As shown in Table 3, and considering the data rates available in IEEE 802.11, a single value of delay ratio is present in each relay area. Delay performance improvement for every direct transmission of k Mbit/s can be expressed as Average Weighted Delay Ratio (AWDR):

$$AWDR_k = \frac{\sum_i \sum_j A_{ijk} \overline{DR}_{ijk}}{\sum_i \sum_j A_{ijk}} \qquad (5)$$

where

$$\overline{DR} = \{DR | DR_{ijk} < 1\} \qquad (6)$$

In addition, we need to define some performance bounds of lower and upper of delay ratio given by

$$\text{Lower Bound of Delay Ratio } (LBDR_k) = \min\{\overline{DR}\} \qquad (7)$$

$$\text{Upper Bound of Delay Ratio } (UBDR_k) = \max\{\overline{DR}\} \qquad (8)$$

As depicted in Table 2, the value of k should be 1 Mbit/s and 2 Mbit/s in IEEE 802.11b to satisfy the $DR_{ijk} < 1$. Similar to delay performance, we can define the metrics of (AWCC) to (UBCC) respectively for average, minimum and maximum of cooperative capacity.

$$AWCC_k = \frac{\sum_i \sum_j A_{ijk} \overline{CC}_{ijk}}{\sum_i \sum_j A_{ijk}} \qquad (9)$$

$$\text{Lower Bound of Cooperative Capacity } (LBCC_k) = \min\{\overline{CC}\} \qquad (10)$$

$$\text{Upper Bound of Cooperative Capacity } (UBCC_k) = \max\{\overline{CC}\} \qquad (11)$$

where

$$\overline{CC} = \left\{ CC | CC_{ijk} = \frac{1}{DR_{ijk}} \text{ and } DR_{ijk} < 1 \right\} \quad (12)$$

In Section 4, we present the relay area based on delay ratio for different revisions of IEEE 802.11. We also consider the average value, lower and upper bound of delay and capacity performance for different direct transmission data rates.

5 Power Performance and Energy Efficiency in Cooperative IEEE 802.11

Energy efficiency in networks using IEEE 802.11 is affected by factors such as the transmit power used and the processing power required for forwarding packets by mobile nodes. Evaluating how throughput varies with the use of relays is important because it allows to also study the resulting energy efficiency. In a non-cooperative direct transmission, the power allocation is carried out only by the source node, while in a cooperative scenario both source and relay nodes should allocate power to complete the transmission. The source node requires power to transmit the packet to the relay, while the relay requires power to forward the packet to the destination. Due to the multi-rate nature of IEEE 802.11, source nodes must collect information from other nodes and reason over it. This way they are able to take an informed decision in whether to send packets through a relay, or directly to the destination. For a direct transmission in the scenario depicted in Figure 4(a), the average received power can be expressed (in dBm) as:

$$P_{r_D} = P_{r_D} + G_t + G_r - PL_d \quad (13)$$

where P_{r_D} (in dBm) is the power radiated by the source in a direct transmission, G_t and G_r are the transmitter and receiver antenna gains, respectively, and PL_d the path loss in dB between source and destination. Considering isotropic antennas, $G_t = G_r = 0$ dBi and the path loss is given by [31]:

$$PL(d) = PL(d_0) + 10n \log_{10} \frac{d}{d_0} \quad (14)$$

where $PL(d_0)$ is the path loss at $d_0 = 1$ m, and $PL(d_0) = -20 \log_{10}(c/4\pi f d_0) = 40.2$ dB at 2.4 GHz, d is the distance between transmitter and receiver and n is the path loss coefficient. For indoor environments with obstructions, such as inside buildings, the path loss coefficient is

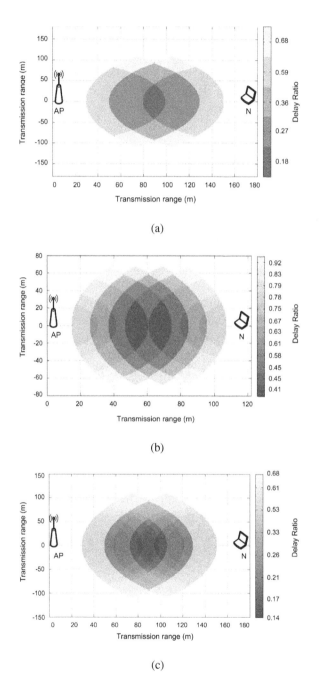

Figure 4 Relay area versus delay ratio for end to end direct transmission of (a) 1 Mbit/s-IEEE 802.11b, (b) 6 Mbit/s-IEEE 802.11g and (c) 1 Mbit/s-IEEE 802.11bg.

between 4 and 6 [31]. The SNR of the received signal for the power noise of N_0 can be expressed as (SNR_d):

$$SNR_d = Pr_D - N_0 = P_{tD} + G_t + G_r - PL_D - N_0 \qquad (15)$$

In a cooperative communication for the same scenario (Figure 4(b)), the average power received by the relay and destination nodes, and the SNR of the received signal can be given by

$$Pr_{rly} = Pt_S + G_t + G_r - PL(d_1) \qquad (16)$$

$$Pr_{Dst} = Pt_R + G_t + G_r - PL(d_2) \qquad (17)$$

$$SNR_{Rly} = P_{Rly} - N_0 = Pt_S + G_t + G_r - PL(d_1) - N_0 \qquad (18)$$

$$SNR_{Dst} = P_{Dst} - N_0 = Pt_R + G_t + G_r - PL(d_2) - N_0 \qquad (19)$$

Obviously, the symmetric cooperative scenario can provide the energy efficiency when $PL(d_1)$ and $PL(d_2)$ are minimum and the conditions of Equation (20) are satisfied.

$$d_1 = d_2 \, and \, Pt_S = Pt_R \qquad (20)$$

and by substituting Equation (20) into (18):

$$Pr_{rly} = Pr_{Dst} = Pt_S + G_t + G_r - PL(d/2) \qquad (21)$$

In order to express the power gain, we define the ΔP as

$$\Delta P = Pr_D - Pr_{Rly} = P_{tD} - Pt_S - PL(d) + PL(d/2) \qquad (22)$$

ΔP can be expressed as

$$\Delta P = SNR_D - SNR_{Rly} \qquad (23)$$

$$\begin{cases} \Delta P = SNR_D - SNR_{Rly} \\ \Delta P = P_{tD} - Pt_S - PL(d) + PL(d/2) \\ \quad = P_{tD} - Pt_S - 10n \log_{10} 2 \\ \quad = P_{tD} - Pt_S - 3n \end{cases} \qquad (24)$$

Substituting (23) into (24) yields

$$SNR_D - SNR_{Rly} = P_{tD} - Pt_S - 3n \qquad (25)$$

The possible values for the medium path loss coefficient n and minimum SNR, required to support the corresponding data rates in IEEE 802.11

Table 4 Data and transmission ranges of IEEE 802.11bg.

IEEE 802.11g	Data rate (Mbit/s)	6	9	12	18	24	36	48	54
	Typical Range (meter)	122	107	96	85	75	61	42	31
	Min-SNR (dB)	8	9	11	13	16	20	24	25
IEEE 802.11b	Data rate (Mbit/s)	1	2	5.5	11				
	Typical Range (meter)	180	150	130	100				
	Min-SNR (dB)	2	2.9	5.4	10				

standard can determine the range of $(Pt_D - Pt_S)$. Let us suppose $SNR_D - SNR_{Rly} = \beta$ and SNR_D and SNR_{Rly} provide the symmetric and delay ratio ≈ 1. Therefore,

$$Pt_D - Pt_S = \beta + 3n \qquad (26)$$

It can be easily concluded that the final power gain is

$$PG = Pt_D - Pt_S - 3 \qquad (27)$$

The value of 3dB is related to symmetric scenario and equal allocated power in source and relay of cooperative scenario. Thus,

$$PG = \beta + 3(n - 1) \qquad (28)$$

In the next section, we will demonstrate how the range of n can determine the minimum and maximum value of PG.

6 Simulation and Results

In order to evaluate the delay performance, capacity performance and energy efficiency of cooperative transmissions as discussed in the previous sections, we devised a scenario with two nodes communicating by using IEEE 802.11bg, and then obtain the maximum distance and the minimum SNR required for maintaining all data rates supported by IEEE 802.11b and IEEE 802.11g. Table 4 indicates the data rates achievable at the different transmission ranges, and the minimum SNR required for $BER < 10^{-5}$: as expected, as the distance between any two nodes increases, the data rates will be adapted down. IEEE 802.11g supports between 54 and 6 Mbit/s, while IEEE 802.11b supports rates between 1 and 11 Mbit/s.

An important aspect of this analysis is that the rates supported are discrete as this number is well known and limited. Therefore, it becomes possible to enumerate all possible cooperation scenarios and evaluate their capacity and energy boundaries. More recent amendments to the protocol provide a greater number of transmission modes, however the principles drawn in this work can be applied to any future amendment. All results below were obtained through the popular simulation tool, OMNET++ and using the Mobility Framework.

6.1 Delay and Capacity Performance

In order to show the relay area versus delay ratio using a geometrical representation, we select just the minimum data rate supported by each standard. This corresponds to the situation showing more possibilities of cooperation, or at higher gain. A graphical representation of the geometries for various relay area in IEEE 802.11b, IEEE 802.11g and IEEE 802.11bg standards is depicted in Figure 4, when the end-to-end data rate is the minimum data rates supported in each standard. The color of the spectrum bar indicates the delay ratio achieved: as the delay ratio increases (lighter color), the performance of the cooperation channel decreases.

As shown in Figure 4, IEEE 802.11b presents values of delay ratio between 0.18 and 0.68, whereas in IEEE 802.11g these values vary from 0.45 to 0.91. Cooperation in IEEE 802.11bg experiences values for delay ratio between 0.14 and 0.68. Therefore, IEEE 802.11bg provides the best potential for cooperative relaying. The variation rate of delay ratio in IEEE 802.11bg is more than IEEE 802.11g and IEEE 802.11b, because of more possibilities for cooperation in IEEE 802.11bg compared to the other ones. This means that for wireless network with mobility scenario, the stability of relay nodes with constant delay ratio in IEEE 802.11bg is less than that in IEEE 802.11b and IEEE 802.11g. Figure 5 presents the average value and lower and upper bounds of delay ratio (i.e., AWDR, LBDR and UBDR) for all direct data rates supported by IEEE 802.11b (g and bg) as discussed in Equations (5) to (8). Figure 5(a) shows the delay ratio of those cooperative scenarios leads to a reduction in delay when using IEEE 802.11b (e.g. 1 or 2 Mbit/s). Beneficial values of the delay ratio ($<$ 1) in IEEE 802.11g can be achieved for direct data rates of 6, 9, 12 and 18 Mbit/s (Figure 5(b)) while in IEEE 802.11bg, cooperative communication can be beneficial for direct data rates of 1, 2, 5.5, 6, 9, 12 and 18 Mbit/s. It is worth mentioning that the value of AWDR of some direct data rates in IEEE 802.11bg is lower than the same data rates in IEEE 802.11b and IEEE 802.11g. As an example, AWDR of 1 Mbit/s varies from 0.44 in IEEE 802.11b to 0.4 in IEEE 802.11bg, and the AWDR of 6 Mbit/s also changes from 0.7 in IEEE 802.11g to 0.65 in IEEE 802.11bg. Thus, cooperation in IEEE 802.11bgcan potentially achieve more performance than when using IEEE 802.11b and IEEE 802.11g in terms of delay reduction, and especially for the similar end-to-end direct data rates due to higher number of situations where cooperation is beneficial.

Figure 6 depicts the cooperative capacity in term of average value, lower and upper bounds in IEEE 802.11b (g and bg) as discussed before in Equations (9) to (12). Figure 6 demonstrates that IEEE 802.11bg has a larger capacity improvement of the cooperative channel in relation to IEEE 802.11b and IEEE 802.11g. As is can be seen in Figure 6, the average weighted cooperative capacity (AWCC) has increased by 18% from IEEE 802.11bg to IEEE 802.11b, for an end-to-end data rate of 1 Mbit/s, while it increases by 40% in IEEE 802.11bg when compared to IEEE 802.11b, and for end-to-end data rate of 2 Mbit/s. This will also provide better energy efficiency, as the distance between source and relay is half of the distance between source and destination, thus reducing the power requirements for a successful transmission.

6.2 Energy Efficiency

To evaluate the energy efficiency achieved by a cooperative scheme, we select an indoor environment with obstructed communication paths (i.e. a normal building with walls and furniture). The path loss coefficient of this environment varies between 4 and 6 [32]. Table 5 indicates the data set rates which provide the delay ratio close to 1, the value of $\beta = SNR_D - SNR_{Rly}$ and the power gain (PG) as discussed in Section 3. Figure 7(a) presents the power gain obtained for minimum and maximum value of path loss coefficient for data rates supported in IEEE 802.11b and IEEE 802.11g while the delay ratio is close to 1 and we have no throughput improvement. As can be seen from Figure 7(a), in cooperative scenarios, if communicating with a lower data rate, we can achieve higher energy efficiency, when compared to the higher data rate. The energy efficiency of IEEE 802.11b in a cooperative scenario outperforms IEEE 802.11g when the main purpose of cooperation is energy saving with no improvement over throughput. This has to do with the communication range provided by the lowest data rate of both protocols. In the case of IEEE 802.11, this range is much higher, thus leading to a more energy efficiency communication.

In order to find the minimum of path loss coefficient which provide the energy efficiency for the data rate set present in Table 5, we solve the following equation:

$$PG = \beta + 3(n - 1) \tag{29}$$

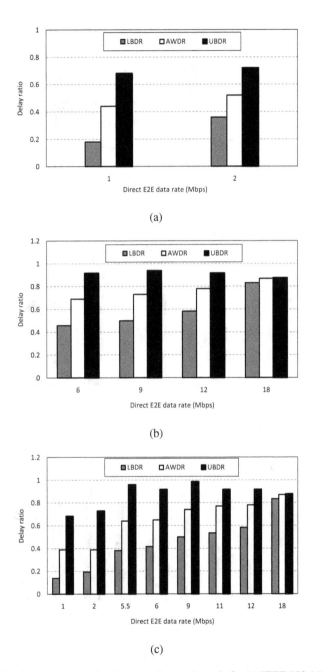

Figure 5 Delay ratio: Average value, lower and upper bounds for (a) IEEE 802.11b, (b) IEEE 802.11g and (c) IEEE 802.11bg.

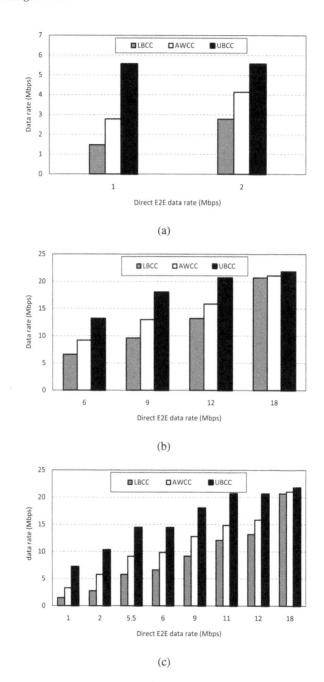

Figure 6 Cooperative capacity: Average value, lower and upper bounds for (a) IEEE 802.11b, (b) IEEE 802.11g and (c) IEEE 802.11bg.

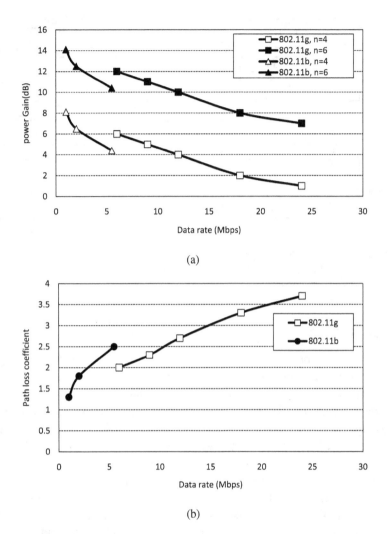

Figure 7 Cooperative capacity: Average value, lower and upper bounds for (a) IEEE 802.11b, (b) IEEE 802.11g.

Figure 7(b) shows the minimum of path loss (n_{min}) for every data rate in IEEE 802.11b and IEEE 802.11g. In order to have energy efficiency through cooperative relay based in IEEE 802.11, the environment with smaller value of path loss coefficient can be more beneficial for low data rates. In addition, Figure 7(b) demonstrates that the wireless environment with path loss coef-

Table 5 Data rate set, β and Power Gain (PG) in IEEE 802.11b and IEEE 802.11g.

| | Data rate (Mbit/s) | Min (SNR) (dB) | Data rate set | | | β (dB) | PG (dB)= Min($n = 4$) $\sim$ Max($n = 6$) |
			SD	SR	RD		
	6	8	–	–	–	–	–
	9	9	–	–	–	–	–
	12	11	6	12	12	-3	6 $\sim$ 12
802.11g	18	13	9	18	18	–4	5 $\sim$11
	24	16	12	24	24	–5	4 $\sim$ 10
	36	20	18	36	36	–7	2 $\sim$ 8
	48	24	24	48	48	–8	1 $\sim$ 7
	1	2	–	–	–	–	–
	2	2.9	1	2	2	–0.9	8.1 $\sim$ 14.1
802.11b	5.5	5.4	2	5.5	5.5	–2.5	6.5 $\sim$ 12.5
	11	10	5.5	11	11	–4.6	4.4 $\sim$ 10.4

ficients of more than 2.5 and 3.7 respectively for IEEE 802.11b and IEEE 802.11g can achieve the power gain in cooperative relay scenarios with data rate set as listed in Table 5.

7 Conclusion

In this work, we presented an architectural reference model for cooperative schemes in wireless cognitive networks, called cooperation loop. According to the capabilities of every wireless networks and cooperation purpose, we can draw a spectrum of features for different phases of cooperation loop. As an example, we discussed the cooperation in IEEE 802.11 standards in term of cooperation loop phases. We also present a theoretical analysis for delay performance and capacity improvement of IEEE 802.11.

Simulation results indicate that IEEE 802.11bg outperforms IEEE 802.11b and IEEE 802.11g in term of throughput due to more possibilities for cooperation. We further discussed how energy efficiency values that can be obtained in cooperative scenarios with single relay node and provided guidelines on the environments beneficial in term of energy saving for cooperative IEEE 802.11 standards. Theses guidelines can be included in cognitive algorithms for cooperation decisions, as discussed in our framework. In the future work, we can consider the cooperative strategy and the imposed overhead of some cooperative protocols.

References

[1] M. Cover and A.A.E. Gamal. Capacity theorems for relay channel. *IEEE Transactions on Information Theory*, 25(5):572-584, 1979.

[2] A. Sendonaris, E. Erkipand, and B. Aazhang. User cooperation diversity, Part I: System description. *IEEE Transactions on Communications*, 51:1927-1938, 2003.

[3] A. Sendonaris, E. Erkip, and B. Aazhang. User cooperation diversity, Part II: Implementation aspects and performance analysis. *IEEE Transactions on Communications*, 51:1939-1948, 2003.

[4] J. Laneman, D. Tse, and G. Wornell. Cooperative diversity in wireless networks: Efficient protocols and outage behavior. *IEEE Transactions on Information Theory*, 50(12):3062-3080, 2004.

[5] P. Liu, Z. Tao, and S. Panwar. A cooperative MAC protocol for wireless local area networks. In *Proceedings of ICC'05*, Vol. 5, pages 2962–1968, 2005.

[6] S. Shankar, C. Chou and M. Ghosh. Cooperative Communication MAC (CMAC) - A new MAC protocol for next generation wireless LANs. In *Proceedings of IEEE International Conference on Wireless Networks, Communications and Mobile Computing*, pages 1–6, 2005.

[7] X. Wang and C. Yang. A MAC protocol supporting cooperative diversity for distributed wireless ad hoc networks. In *Proceedings of IEEE PIMRC'05*, 2005.

[8] A. Sadek, K.J. Ray Liu, and A. Ephremides. Collaborative multiple-access protocols for wireless networks. In *Proceedings of ICC'06*, 2006.

[9] P. Pathak and R. Dutta. A survey of network design problems and joint design approaches in wireless mesh networks. *IEEE Communication Surveys Tutorials*, 2010.

[10] P. Liu, Z. Tao, Z. Lin, E. Erkip, and S. Panwar. Cooperative wireless communications: A cross-layer approach. *IEEE Wireless Communication Journals*, 13(4):84-92, 2006.

[11] H. Shan, W. Zhuang, and Z. Wang. Distributed cooperative MAC for multihop wireless networks. *IEEE Communication Magazine*, 47(2):126-133, 2009.

[12] C. Cetinkaya and F. Orsun. Cooperative medium access protocol for dense wireless networks. In *Proceedings of Third Annual Mediterranean Ad Hoc Networking Workshop (Med Hoc Net)*, pages 197-207,2004.

[13] IEEE, Part 11: Wireless LAN Medium Access Control (MAC)and Physical Layer (PHY) Specifcations, 1999.

[14] M. Heusse, F. Rousseau, G.B. Sabbatel, and A. Duda. Performance anomaly of IEEE 802.11b. In *Proceedings of IEEE Infocom*, Vol. 2, pages 836–843, 2003.

[15] IEEE Std. IEEE 802.11g, Supplement to Part 11: Wireless LAN Medium Access Control (MAC) and Physical Layer (PHY) specifications; Further high-speed physical layer extension in the 2.4 GHz band. 2003.

[16] IEEE Std. IEEE 802.11n, Specific Requirements Part 11: Wireless LAN Medium Access Control (MAC) and Physical Layer (PHY) Specifications amendment 5: Enhancements for higher throughput, 2009.

[17] P. Liu, Z. Tao, S. Narayanan, T. Korakis, and S.S. Panwar. CoopMAC: A cooperative MAC for wireless LANs. *IEEE Journal on Selected Areas in Communications*, 25(2):340-354, 2007.

[18] H. Zhu and G. Cao. rDCF: A relay-enabled mediumaccess control protocol for wireless Ad Hoc networks. *IEEE Transactions on Mobile Computing*, 5(9):1201-1214, 2006.

[19] F. Liu, T. Korakis, Z. Tao, and S.S. Panwar. A MAC-PHY cross-layer protocol for ad hoc wireless networks. In *Proceedings of IEEE WCNC'08*, pages 1792-1797, 2008.

[20] F. Verde, T. Korakis, E. Erkip, and A. Scaglione. On avoiding collisions and promoting cooperation: Catching two birds with one stone. In *Proceedings of IEEE SPAWC*, pages 431-435, 2008.

[21] P. Liu, Y. Liu, T. Korakis, A. Scaglione, E. Erkip, and S.S. Panwar. Cooperative MAC for rate adaptive randomized distributed space-time coding. In *Proceedings of IEEE Globecom'08*, pages 4997-5002, 2008.

[22] P. Liu, C. Nie, E. Erkip, and S. Panwar. Robust cooperative relaying in a wireless LAN: Cross-layer design and performance analysis. In *Proceedings of IEEE Globecom'09*, 2009.

[23] F. Verde, T. Korakis, E. Erkip, and A. Scaglione. A simple recruitment scheme of multiple nodes for cooperative MAC. *IEEE Transaction Communication*, 58(9):2667-2682, 2010.

[24] B. Zhao and M. Valenti. Practical relay networks: A generalization of hybrid ARQ. *IEEE Journal of Selected Areas Communication*, 23(1):7-18, 2005.

[25] S. Cui, A. Goldsmith, and A. Bahai. Energy-effciency of MIMO and cooperative MIMO techniques in sensor networks. *IEEE Journal on Selected Areas Communication*, 22(6):1089-1098, 2004.

[26] R. Ahmad, F.-C. Zheng, and M. Drieberg. Modeling energy consumption of dual-hop relay based MAC protocols in ad hoc networks. In *Proceedings of IEEE ISWPC'09*, 2009.

[27] R. Ahmad, F.-C. Zheng, and M. Drieberg. Delay analysis of enhanced relay-enabled distributed coordination function. In *Proceedings of IEEE VTC'10*, 2010.

[28] J. Kephart and D. Chess. The vision of autonomic computing. *IEEE Computer Journal*, 36(1):41-50, 2003.

[29] J. Mitola. Cognitive radio – An integrated agent architecture for software defined radio. Ph.D. Thesis, Royal Institute of Technology, Kista, Sweden, 2000.

[30] B. Cetin. Opportunistic relay protocol for IEEE 802.11 WLANs. Master Thesis, Royal Institute of Technology, 2006.

[31] T. Rappaport *Wireless Communications: Principles and Practice*. Prentice Hall, 1999.

[32] J. Alonso-Zárate, E. Kartsakli C. Verikoukis, and L. Alonso. Persistent RCSMA: A MAC protocol for a distributed cooperative ARQ scheme in wireless networks. *EURASIP Journal on Advanced Signal Processing*, Special Issue on Wireless Cooperative Networks, 2008.

[33] Rasool Sadeghi, João Paulo Barraca, and Rui L. Aguiar. Metrics for optimal relay selection in cooperative wireless networks. In *Proceedings of IEEE PIMRC'11*, 2011.

[34] Muharrem Ali Tunc. Service differentiation via cooperative MAC protocol (SD-MAC). Master's Thesis, Wichita State University, 2003.

[35] J.S. Pathmasuntharam, A. Das, and K. Gupta. Efficient multi-rate relaying (EMR) MAC protocol for ad hoc networks. In *Proceedings of IEEE ICC'05*, pages 2947-2951, 2005.

Biographies

Rasool Sadeghi is a Ph.D. student in telecommunication engineering, at the MAP-tele program at the University of Aveiro, Portugal. He is also a research member at Instituto de Telecomunicações (IT), Aveiro, Portugal. He has a Master's degree (2004) in Electrical Engineering from Shiraz University, Iran. His current research interests include cooperative and cognitive protocols for MAC protocols in wireless networks.

João Paulo Barraca received the Licenciatura degree in Computers and Telematics Engineering from the University of Aveiro, Portugal in 2004, and two years later a Masters degree in Electronics and Telecommunication Engineering from the same university. He joined Instituto de Telecomunicações in 2003, were he develops his research activities, and in 2008 he became an Invited Lecturer at the University of Aveiro. In 2007 he started pursuing a Ph.D. in Computer Science. His research activities include management mechanisms for community-oriented autonomic networks, and development of experimentation facilities for wireless environments. He has coauthored more than 20 publications, published in journals and conference proceedings.

Rui L. Aguiar received a Ph.D. degree in electrical engineering in 2001 from the University of Aveiro, Portugal. He is currently a professor at the University of Aveiro and an adjunct professor at the INI, Carnegie Mellon University. He is leading a research team at the Institute of Telecommunications, Aveiro, on next-generation network architectures and systems. His participation in European cooperative research is extensive. His current research interests are centered on the implementation of advanced new networks and systems with special emphasis on QoS and mobility aspects. He is a member of ACM and a senior member of IEEE. He has more than 250 published papers in those areas. He has served as technical and general chair of several conferences, such as ICNS'05, ICT'06, ISCC.

Spectrum Utilization and Congestion of IEEE 802.11 Networks in the 2.4 GHz ISM Band

Jan-Willem van Bloem[1], Roel Schiphorst[1], Taco Kluwer[2]
and Cornelis H. Slump[1]

[1]University of Twente, PO Box 217, 7500 AE Enschede, The Netherlands;
e-mail: {j.w.h.vanbloem, r.schiphorst, c.h.slump}@ewi.utwente.nl
[2]Radiocommunications Agency Netherlands, PO Box 450, 9700 AL Groningen,
The Netherlands; e-mail: taco.kluwer@agentschaptelecom.nl

Received 16 October 2011; Accepted: 11 May 2012

Abstract

Wi-Fi technology, plays a major role in society thanks to its widespread
availability, ease of use and low cost. To assure its long term viability in
terms of capacity and ability to share the spectrum efficiently, it is of para-
mount to study the spectrum utilization and congestion mechanisms in live
environments. In this paper the service level in the 2.4 GHz ISM band is
investigated with focus on todays IEEE 802.11 WLAN systems with support
for the 802.11e extension. Here service level means the overall Quality of
Service (QoS), i.e. can all devices fulfill their communication needs? A cross-
layer approach is used, since the service level can be measured at several
levels of the protocol stack. The focus is on monitoring at both the Physical
(PHY) and the Medium Access Control (MAC) link layer simultaneously
by performing respectively power measurements with a spectrum analyzer
to assess spectrum utilization and packet sniffing to measure the congestion.
Compared to traditional QoS analysis in 802.11 networks, packet sniffing
allows to study the occurring congestion mechanisms more thoroughly. The
monitoring is applied for the following two cases. First the influence of
interference between WLAN networks sharing the same radio channel is

Journal of Green Engineering, Vol. 2, 401–430.

investigated in a controlled environment. It turns out that retry rate, Clear-To-Send (CTS), Request-To-Send (RTS) and (Block) Acknowledgment (ACK) frames can be used to identify congestion, whereas the spectrum analyzer is employed to identify the source of interference. Secondly, live measurements are performed at three locations to identify this type of interference in real-live situations. Results show inefficient use of the wireless medium in certain scenarios, due to a large portion of management and control frames compared to data content frames (i.e. only 21% of the frames is identified as data frames).

Keywords: interference, IEEE 802.11e, ISM band, congestion, cross-layer, spectrum sensing.

1 Introduction

Commodity wireless broadband, e.g., Wi-Fi technology, plays a major role in society thanks to its widespread availability, ease of use and low cost. Many new applications have emerged for such technologies, intelligent transportation systems (ITS), Dynamic Spectrum Access (DSA) systems and offloading of traffic from cellular networks. Because the role of Wi-Fi technology is becoming increasingly important, its long term viability in terms of capacity and ability to share spectrum efficiently, must be assured. Current practice shows already that Wi-Fi technology is not very efficient and there is a risk of collapse in certain scenarios, like high density video and data off-load from cellular networks.

In this paper we focus on the crowded 2.4 GHz ISM band. The number of wireless devices (smartphones, laptops, sensors) that use this band is rapidly increasing. In many urban areas not only many WLAN networks can be found, also other systems like Bluetooth, Zigbee and wireless A/V transmission systems use this band. On the other hand there is only a limited amount of spectrum available. So it is very likely that interference between systems in this band will occur. Due to the rapid increase of wireless devices, interference is expected to become even more important. In this paper we address this issue by providing a setup to measure the service level - i.e. can all devices fulfill their communication needs - in this band with focus on WLAN standard IEEE 802.11e. The upcoming IEEE 802.11e is an extension of the 802.11 Wireless Local Area Network (WLAN) standard and is developed to enhance Quality of Service (QoS) support. Furthermore, note that IEEE 802.11e has been incorporated into the current published IEEE 802.11

standard to which we refer - with slight abuse of notation - as IEEE 802.11e. On the other hand, we refer to the traditional IEEE 802.11 technology without QoS support as legacy IEEE 802.11 systems.

The service level measurements can be split into two parts: utilization and congestion. Utilization or spectrum sensing means how much of the 2.4 GHz ISM band is in use for a certain area. When the utilization has been measured, it is still unknown what the quality is of the wireless connections i.e. degradation cannot be measured; a measure for quality is congestion. Basically it analyzes the frame types transmitted through the wireless medium. Congestion occurs if the retry frame rate or the number of control frames is high. WLAN systems often use these frames to mitigate interference from other WLAN networks. Measuring the service level is an crucial parameter to assess the efficiency of the WLAN networks on other wireless networks and to identify co-existing issues. For the co-existence of competitive 802.11e networks the available resources have to be shared by overlapping and competing WLANs with high offered traffic load. Measuring and assessing the QoS of competing overlapping 802.11e compatible WLANs is the main topic of this paper.

The paper is organized as follows. First in section 2 the related work on monitoring QoS is highlighted. Next, in Section 3 the 802.11(e) co-existence mechanisms are presented. In Section 4 the influence of overlapping 802.11e WLANs on the same radio communication channel is investigated based on measurements set up in a controlled environment. In Section 5 the monitoring results, taken in real environments without control, are presented. Section 6 concludes the paper.

2 Related Work on Monitoring QoS

The OSI protocol stack of the IEEE 802 standard consists of five layers where the lower two layers involve the wireless aspects of communication, i.e. the Physical (PHY) layer and the Medium Access Control (MAC) link layer. To assess the overall Wi-Fi QoS the monitoring must take place at these lower two layers. At the PHY layer this entails a threshold-based approach to determine spectrum occupation. This value can be used to assess utilization of the band. On the other hand, at the MAC link layer the total frame rate and retry frame rate have been found to be good parameters for quantifying the level of utilization and network degradation [37]. In addition, monitoring of the proportions of the different MAC link frame types can be used to assess the overhead in a Wi-Fi network, i.e., data, management and control (e.g.

Clear-to-Send(CTS), Request-to-Send (RTS), and (Block) Acknowledgment (ACK) frames. Much of the papers discuss the QoS on the MAC link layer only, i.e. in [18, 19] the performance evaluation of 802.11 WLAN systems is provided and in [7, 25, 27, 28] the 802.11e MAC link layer QoS. The advantage herein lies in the broad range of MAC link layer specific aspects and protocols which can be modeled and compared. Notice that in this paper we use both RF spectrum and MAC link layer monitoring in order to address the overall QoS.

Not much related literature has been found on overall QoS. First, a description of the overall Wi-Fi QoS monitoring is provided in *Estimating the Utilization of Key License-Exempt Spectrum Bands*, Final report, issue 3, April 2009 by Mass Consultants Limited commissioned by the British regulator Ofcom [37]; here monitoring takes place at various locations in the UK and live readings are obtained. In the latter report the QoS is assessed at the MAC link layer level but is only monitored at the highest abstraction level, i.e., whether the type of frame is data, management or control; this means that no deep packet inspection is carried out (e.g. ACK, RTS, CTS, etc) so that a description of the occurring mechanisms cannot be provided thoroughly. Furthermore, in [34] the total QoS is assessed of WLAN systems in live environments using a cognitive radio approach. Regarding monitoring QoS, the impact of overlapping WLAN interference is not addressed in the above-provided papers.

2.1 QoS: Overlapping WLANs

The problem of overlapping WLANs is for legacy 802.11 systems mainly focused on co-channel interference. Note that in this paper we do not cover adjacent channel interference (e.g. see [26, 36]) for the sake of brevity. Elaborating on co-channel interference simulation based studies – using game theory models – are provided e.g. in [18, 23], and experimental studies in [4, 12, 13]. This entails MAC link layer performance models for assessing QoS; mainly the 802.11 interference impact on concurrent data throughput is investigated. For instance, it is shown in [4] that the throughput remained nearly constant irrespective of the distance between the 802.11 Access Points (APs), indicating that distance does not not have an effect on the throughput as long as the APs are in the carrier sensing range. In addition, in [13] the performance of a particular AP in a highly congested environment is evaluated, i.e. many concurrent WLAN networks. Here the live recordings show that in case of bad link quality the AP becomes inactive for a while, i.e. the AP operates in cycles of respectively activity and inactivity. Furthermore, the related problem of 802.11 hidden node issues and the impact of the RTS/CTS mechanism are addressed in [14]. However, a drawback of the above-mentioned studies is that 802.11e is not taken into account.

Regarding 802.11e, stochastic simulations on overlapping 802.11e WLANs are provided in [21, 22, 24, 25]. This entails a 802.11e MAC link layer investigation using respectively the aggregate data throughput and data packet delay as performance metrics. Another simulation study on the related problem of hidden node terminals in the vicinity of 802.11e WLAN is given in [16]. Note that the advantage of [16,21,22,24,25] lies in the thorough theoretical research of overlapping 802.11e networks. However, the experimental work and the overall QoS are beyond the scope of their work.

2.2 Contributions

To assess the overall QoS for 802.11 systems – in highly congested environments with overlapping WLANs – an experimental validation is required. In line with this, the main contributions of this paper are:

- A measurement setup to measure the congestion in the 2.4 GHz ISM band. The setup is able to analyze three WLAN channels in parallel. In addition, the setup allows to identify which frame types (and sub fields) are transmitted, useful to identifying congestion (This setup is an extension of the work by Mass Consultants [37]).
- Reporting the results of measurements in a controlled environment where the interference of a second WLAN network is measured. From these results one can conclude that the RTS/CTS mechanism found in the WLAN standard (used for the hidden node problem) is one of the main sources of congestion. WLAN networks often identify interference as a hidden node.
- Reporting the experimental results of devices that use the new 802.11e extension.
- Reporting the results of measurements in a live environment (college room, office room, city center). The results reveal that for a college room only 21% are actual data type frames. Almost 70% are control frames.

3 IEEE 802.11 Co-existence Mechanisms

In this section a short overview of legacy 802.11 mechanisms is provided to access the shared medium; subsequently the mechanisms of its enhanced

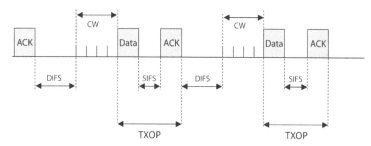

Figure 1 The 802.11 DCF protocol timeline. For each data frame transmission a back-off period precedes. The bounded time interval reserved for transmission is denoted as Transmission Opportunity (TXOP) using the 802.11e terminology.

version 802.11e are highlighted. It is important to know these co-existence mechanisms to analyze the experimental results.

3.1 Legacy IEEE 802.11

The legacy 802.11 WLAN technology is characterized by its best-effort service (no guarantee of any service level to users/applications). In legacy 802.11 WLAN the PHY layer comprises 11 channels in the 2.4 GHz ISM band where each channel has a bandwidth of 20 MHz.

On the other hand, legacy 802.11 MAC link layer defines two procedures for 802.11 stations to share a common radio channel [3], i.e., Distributed Coordination Function (DCF) and Point Coordination Function (PCF). Here the mandatory DCF is based on Carrier Sense Multiple Access with Collision Avoidance (CSMA/CA) and operates in a listen before talk fashion; DCF is designed for time-bounded services. Compared to DCF, the optional PCF is a contention-free scheme using a central controller to schedule channel access [5]. Note, in the sequel only DCF is highlighted as most of the current 802.11 technologies operate in DCF mode only. The timing of the DCF scheme is depicted in Figure 1. Essential for DCF is the so-called Contention Window (CW) which is maintained at each station. Based on the size of the CW a Back-off Counter (BC) is determined as a random integer drawn from an uniform distribution over the interval $[0, CW]$. Note, a station is allowed to transmit a frame if the channel stays idle during the so-called DCF Interframe Space (DIFS) time interval and if it subsequently remains idle during the followed-up back-off process. In addition, for each successful frame reception the receiving station immediately sends ACK frame. Note that the ACK frame is transmitted after a Short IFS (SIFS), which is shorter

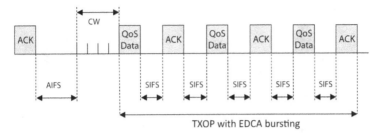

Figure 2 The 802.11e EDCA timeline. Using the 802.11e terminology, the MAC data units are denoted as QoS data frames.

than the DIFS, i.e. $DIFS = SIFS + 2*Slottime$. To mitigate congestion the following mechanisms occur as part of DCF: a transmission failure causes the CW to increase in an exponential fashion; the other way around, the CW is decremented linearly in case of successful transmission (i.e. ACK reception).

3.2 IEEE 802.11e

Occurring 802.11e Mechanisms without Communication Errors

To support QoS the IEEE 802.11 Task Group E has defined 802.11e [2, 30] as an enhancement to the above-described legacy 802.11 technology. At the PHY layer the 802.11e standards allows 20/40 MHz mode protection. This entails automatic channel switching possibilities in order to be compliant with legacy 802.11 systems [6]. However, a maximum bandwidth width of respectively 22 and 42 MHz is tolerated due to the extra 1MHz on each end that the channel is allowed to attenuate. At the 802.11e MAC link layer the Enhanced Distributed Channel Access (EDCA) protocol is defined to provide priority-based distributed channel access [6]. The EDCA is controlled by the HC (Hybrid Coordinator) – resided in the AP – and is developed to be compatible with the legacy 802.11 MAC protocol. Note, in literature [7] EDCA is also referred to as Enhanced (DCF), i.e. EDCF; moreover, according to the 802.11e terminology a data packet needs to be denoted as *QoS Data*. Under the EDCA, each channel access does not result necessarily in a single data frame transmission but more data frames are allowed in burst transmissions [31]. This means that a particular station – with gained channel access – has the right to initiate transmissions during the granted Transmission Opportunity (TXOP) time interval [8, 15, 20, 33]. In addition, another 802.11e parameter is defined: Arbitration Interframe Space (AIFS). Here AIFS is the legacy 802.11 DIFS equivalent, but its size depends on the frames/packet

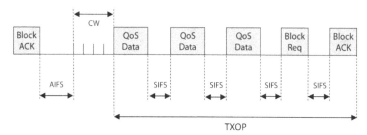

Figure 3 The IEEE 802.11e Wireless LAN with Block Acknowledgments. After a burst of consecutive data frames a Block Request frame is sent by the transmitter which in turn is acknowledged with a Block ACK frame by the receiver.

priority (note that AIFS is at least DIFS). Furthermore, in EDCA-TXOP mode the CW size is made significantly smaller than that for 802.11 DCF. The timeline of 802.11e communication in EDCA-TXOP mode is depicted in Figure 2. As a consequence, the higher throughput of EDCA-TXOP can cause jamming interference to other systems operating in the 2.4 GHz ISM band [35].

In highly congested wireless environments the 802.11e devices are allowed to switch transmission mode in order to increase throughput and reduce overhead, but this is not obligatory [11, 17, 29]. This is achieved by acknowledging several consecutive data frames with one response, i.e. a block ACK. The decision on the mode of transmission is made by the AP, and this can be set up within an ongoing TXOP transmission. The 802.11e block ACK feature is explained in Figure 3.

Occurring 802.11e Mechanisms with Communication Errors

The operation of 802.11e – with RTS/CTS protection enabled – in response to channel errors is the following. First the data frame losses are discussed which is depicted in Figure 4(a). A possible failed data transmission is detected by the sender due to the ACK timeout which in turn triggers a new back-off period. If the channel is sensed idle a RTS/CTS exchange is initiated which is followed up by the actual data retransmission. However, in a highly congested environment a subsequent RTS packet collision loss is not unlikely. The sender detects this by means of the CTS timeout mechanism so that another RTS packet can be queued for transmission. This in turn can lead to a (large) number of consecutive RTS frames transmitted over the channel as illustrated in Figure 4(b).

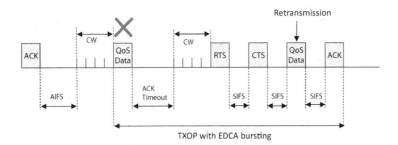

(a) The data frame loss, detected by the ACK timeout at the transmitter.

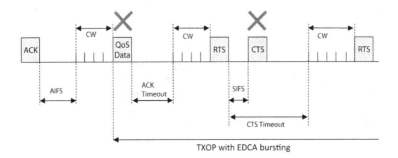

(b) The RTS packet loss, detected by the CTS timeout mechanism. In a highly congested environment this can lead to a large number of consecutive RTS frames.

Figure 4 The 802.11e features in conjunction with RTS/CTS to cope with channel errors.

4 Interference Mechanisms

The interference between 802.11e WLANs sharing the same radio channel is highlighted in this section. This is important before measurements in a live environment can be made in order to identify the interference mechanisms better from the live recordings. Interference mechanisms are investigated by performing measurements in a controlled environment. In order to determine the mechanisms that lay behind this type of interference a cross-layer approach is introduced comprising monitoring at both the PHY and MAC link layer.

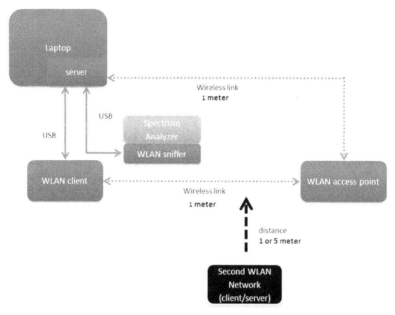

Figure 5 Measurement setup to assess the influence of an interfering network in a controlled environment.

4.1 Setup

The measurement setup consists of two 802.11e WLANs with network load, respectively denoted as the main network (first network) and the interfering network (second network). The measurements are set up in a controlled environment, i.e. a quiet environment without any nearby devices active in the 2.4 GHz ISM band.

To assess the interference influence of the second network on the main network RF monitoring and MAC link layer packet capturing take place simultaneously (see Figure 5). In this setup we use passive methods for monitoring, i.e. the measurement setup only receives signals. The two networks are characterized in general terms as follows:

1. Main 802.11e WLAN: a client streaming data from a server system at a fixed rate via the AP of network 1.
2. Interfering 802.11e WLAN network: positioned in range of the main network; data transmission takes place between a client at server through the AP of network 2 on the same WLAN channel as the main network.

The measurement equipment is located in the main network close the client/server devices, since the influence of the interfering network on the main network needs to be assessed. The measurement equipment consists of:

- Packet sniffer: to capture raw packets on MAC link layer level.
- RF monitoring equipment: spectrum analyzer tuned to the 2.4 GHz ISM band.

Furthermore, to assess the level of interference certain performance metrics are required to determine service level degradation. At the MAC link layer level the following metrics are used: packet frame rate, retry rate (only defined for data and management frames), and subfields (e.g. RTS/CTS/ACK, etc.).

In addition the measurement parameters associated to the setup are listed below:

- In the WLAN network the client, AP and server are within a radius of 1 meter.
- The AP operates at a maximum rate of 300 MBits/sec (IEEE 802.11e mode) with 20/40 MHz protection mode enabled.
- The WLAN channel is set to channel 11: 2451–2473 MHz.
- Spectrum Analyzer: CRFS RFeye equipment [1].
- The field strengths are measured and depicted in dBm. The spectrum analyzer measures the spectrum every 200 ms at a frequency resolution of 4 kHz. For each frequency bin the instantaneous peak power is recorded.
- The packet sniffing software is custom-made, tailored to the specific monitoring needs.
- The sniffer application filters the frames by (destination) MAC address. So it allows to measure both the frames transmitted by the client and server.
- Data is transferred using the Unix network tool Iperf, and a session is set up between server and client devices.

Note that in this setup both server and client are connected to one Laptop using virtualization software. In Iperf the UDP (User Datagram Protocol) mode has been selected instead of TCP, because it allows to study the WLAN interference better. The reason for this is that TCP is supplemented with control algorithms that set back the frame rate to a lower level when the packet loss increases. This is not the case regarding the UDP mode – which is connectionless – and therefore gives a better understanding of the involved interference mechanisms.

4.2 Experiments

Using the measurement setup of Figure 5 two experiments are carried out. The conducted experiments are set up in order to investigate the impact of distance between two overlapping WLANs w.r.t. the service level. Here, in both 802.11e WLANs the QoS data rate between server and client is fixed to 170 frames/sec.

- Experiment 1: the interferer 802.11e WLAN at 1 m distance.
- Experiment 2: the interferer 802.11e WLAN at 5 m distance.

For both experiments the interferer traffic load is high, i.e. 424 frames/sec. In perspective, the maximum packet rate for IEEE 802.11g is around 2200 frames/sec.

4.3 Results

4.3.1 Experiment 1: WLAN Interferer Network at 1 m

Regarding PHY layer, the monitoring the results are depicted in Figure 6(a) (RF spectrum) and Figure 6(b) (RF occupancy). On the other hand, the MAC link layer monitoring results are shown in respectively Figure 7(a) (transmitted frames with destination client), Figure 7(b) (transmitted frames with destination server), and Figure 7(c) (transmitted frames from the interfering 802.11e WLAN). During the experiments no parameters have been changed, so the figures show behavior in time. Note that the that the occupancy has been determined by placing the threshold several dBs above the noise floor at −155 dBm.

- 0–20 seconds: the server data rate is at a normal level, i.e. 170 frames/sec. This also holds for the client with a similar ACK frame rate. The retry rate is nearly zero. The amount of QoS data load in the interferer network is very low.
- 20–80 seconds: period of congestion. The QoS data rate at the server drops gradually from 170 frames/sec to 50 frames/sec. The RF activity increases to occupancy values above 50%. In addition, there is an increase in the number of retry frames in conjunction with a very strong drop of detected ACK frames to nearly zero. Besides, no RTS/CTS frames have been injected in the wireless medium by the server of the

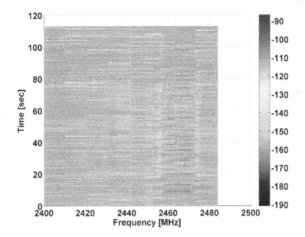

(a) Spectrum analyzer RF recordings (field strength values are in dBm)

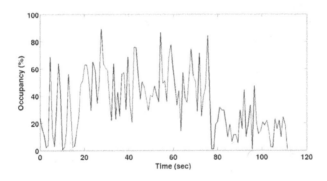

(b) The occupancy of WLAN channel 11

Figure 6 The RF spectrum monitoring results for experiment 1 with nearby interfering 802.11e WLAN at 1 m. A rise in RF field strengths is visible during the period of high interferer activity, i.e. 20–80 seconds. Moreover, the extra bandwidth channel is visible in the spectrogram due to the 20/40 MHz protection mechanism.

main network. However, a rise of RTS/CTS activity is measured coming from the interferer network (see Figure 7(c)). Thus it is likely that the interferer AP has initiated the RTS/CTS mechanism in order to mitigate interference; this according to the described 802.11 co-existence mechanisms in section 3, i.e. the main 802.11e WLAN network has been identified as hidden node. In a similar fashion an increase of (duplicate)

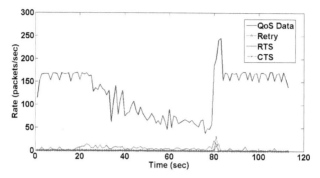

(a) Packets with destination client.

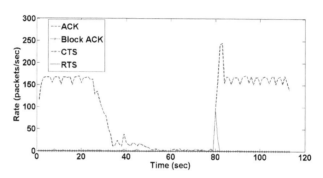

(b) Packets with destination server.

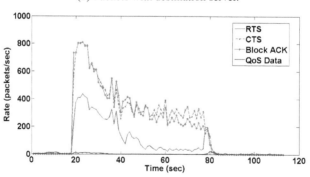

(c) Packets with a destination in the interferer network.

Figure 7 The MAC link layer monitoring results from experiment 1 with the interfering 802.11e WLAN at 1 m. The impact of the interference network is visible mainly in the period of 20–80 seconds. Here the CTS and Block ACK frames in the interferer network are counted twice by the sniffer application because these frames are retransmitted by the AP and are not filtered as such.

802.11e Block ACK frames is measured. However, the amount of QoS data load in the interferer 802.11e WLAN remains nearly zero.

- 80–110 seconds: this period starts with a short peak (at 80 seconds) of QoS data in the main 802.11e WLAN which sets back to a rate of 170 frames/sec subsequently. In addition, the retry rate falls back to a minimum together with a low RF activity level of around 35%. The QoS data rate of the interferer WLAN remains very low.

From the above-presented figures it can be concluded that the co-channel interference of a nearby second 802.11e WLAN network (at 1 m) occurs in cycles of respectively activity and inactivity. This is in accordance with the experimental live results in [13] where the impact of overlapping WLANs on the aggregate AP throughput is recorded. During the active period this entails an initial strong increase of RTS/CTS frames. Next, the number of RTS/CTS frames falls back exponentially during this interval. This can be explained by the high number of occurring collisions with frames coming from the main network which in turn triggers the CSMA/CA mechanism of the interferer network to back off. However, the amount of traffic load from the 802.11e interferer 802.11e WLAN is very high and has a jamming character which can be explained by the shorter CW interval due to the EDCA-TXOP transmission mode of 802.11e [2]. This is in line with the observations in [35] where the jamming impact of 802.11e is observed.

During this period of congestion a strong increase of RF occupancy is observed in the spectrum occupancy plot in Figure 6(b) in conjunction with the use of an extra channel which leads to a total bandwidth of 40 MHz; this in contrast to the single 20 MHz channel mode employed in the cycle of interferer inactivity. However, according to [6] the 20/40 MHz protection mechanism has been developed to operate differently, i.e. switch to single channel 20 MHz mode in case of detected interference. In addition, there seems to be a second mechanism active, i.e. in the main network the retry rate increases due to the high load of RTS, CTS and Block ACK frames coming from the interfering network. This in turn triggers the the CSMA/CA mechanism to back off which leads to a decline of QoS data injected by the server of the main network (see Section 3 describing the co-existence mechanisms).

Finally, it is observed that a period of high congestion can be identified by the high amount of respectively RTS, CTS and duplicate Block ACK frames. It turns out based on this experiment that Block ACK frames are a sign of congestion for 802.11e systems.

4.3.2 Experiment 2: 802.11e WLAN Interferer Network at 5 m Distance

In this experiment the 802.11e WLAN interferer network is positioned at a larger distance from the main network, i.e. at 5 m distance. The configuration settings are similar to those in experiment 1. First, the PHY monitoring results are depicted in respectively Figure 8(a) and Figure 8(b), i.e. the RF spectrum and the spectrum occupancy of channel 11. Second, the MAC link layer monitoring results are shown in Figure 9(a) (packets with destination client),

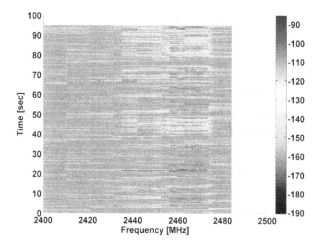

(a) Spectrum analyzer RF recordings (field strength values are in dBm)

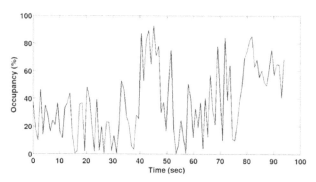

(b) The occupancy of WLAN channel 11

Figure 8 Experiment 2: the RF spectrum monitoring results with interfering 802.11e WLAN at 5 m. Two periods of congestion are visible, i.e. between 40–50 seconds and 80–95 seconds, by the higher level of RF field strength values.

Figure 9(b) (packets with destination server), and in Figure 9(c) (packets with a destination in the interferer network).

- 0–40 seconds: the transmitted frame rate at the server is around 170 frames/sec in normal transmission mode; the same holds for the client rate (ACK frames). The average RF occupancy is below 40%. The packet loss in the network, i.e. retry rate, is negligible low.
- 40–50 seconds: the first period of congestion. A rise of RTS, CTS and Block ACK frames from the interferer 802.11e WLAN. At the start of this congestion period there is a peak in the number of RTS frames at the server. Subsequently the QoS data rate in the main network drops in conjunction with the ACK rate at the client site. In addition, the retry rate remains very low. Regarding the PHY layer, the RF activity is very high (> 80%) and an extra 20 MHz channel is visible in the spectrogram of Figure 8(a).
- From 50–80 seconds: QoS data frame rate at normal level in the main network; same holds for the ACKs at the client site. The number of RTS, CTS and Block ACK frames drops - compared to the preceding period of congestion - but is still present at a significant high level (peak rates: 150-200 frames/sec). In addition, it is observed from the recorded data readings that the QoS data - injected by the server of the interfering 802.11e WLAN - occur in EDCA-TXOP block burst mode. Regarding PHY layer, there is a decline in RF activity but the occupancy level is significantly higher compared to the first period of congestion.
- From 80–95 seconds: the second period of congestion. A rise in the number of RTS, CTS and Block ACK frames occurs in the interferer 802.11e WLAN up till a level of 500 frames/sec. However, this is half the amount compared to the first period of congestion. On the other hand, the decline in QoS data and ACK frames is more significant compared to the first period of congestion. The RF occupancy is in the range of {50–80%}, i.e. significantly lower compared to the first congestion period.

In general, the impact of an overlapping co-channel 802.11e WLAN at 5m distance is comparable to the 1 m case. Similarly, the interference network shows cycles of respectively activity and inactivity as stated in [13] for the aggregate AP throughput. During these active cycles the service level of the main network degrades to a lower level, i.e. the QoS data rate drops by half the amount. This is in line with [4] where experiments show that the impact on the aggregate throughput is irrespective of distance as long as the WLANs are in carrier sense range of each other.

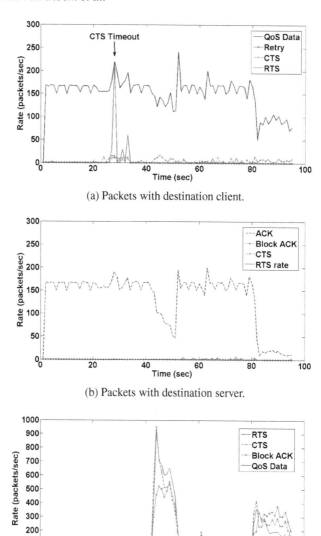

(a) Packets with destination client.

(b) Packets with destination server.

(c) Packets with a destination in the interferer network.

Figure 9 The MAC link layer monitoring results from experiment 2 with the interfering 802.11e WLAN at 5 m. The impact of the interference network is visible mainly in the periods 40–50 seconds and 80–95 seconds. Here the CTS and Block ACK frames in the interferer network are counted twice by the sniffer application because these frames are retransmitted by the AP and are not filtered as such.

In accordance with the analysis of co-existence mechanisms in Section 3, the experimental results show the CTS timeout mechanism which occurs at the start of the congestion period as depicted in Figure 9(a).

According to the experiments the influence of interferer distance on the service level is limited but still perceptible. Key is that due to to higher separation distance, i.e. 5 m instead of 1 m, the channel is more often sensed idle. This means that more timeslots are available to initiate transmissions for the interferer 802.11e WLAN. This is visible in the period {50–80} seconds where the level of RTS/CTS frames – injected in the network by the interferer 802.11e WLAN – is higher compared to the 1 m case. In addition, to capture channel access the 802.11e block burst transmission mechanism becomes active at the interferer AP. Note that the moderate level of RTS/CTS packets from the interferer does not harm the QoS data rate during this interval. However, the moderate RTS/CTS load causes the CW to be adjusted to a larger size for both WLANs. As a consequence, in the subsequent period of congestion the interferer network injects RTS and CTS packets at a lower rate into the wireless medium. Note that this is also the case for the amount of QoS data injected by the server of the main network, i.e. visible as strong steep drop of QoS data in Figure 9(a).

5 Live Measurements

In this section the setup and the results are presented of measurements in a live environment.

5.1 Setup

A modified measurement setup has been used compared to the setup employed in the interferer measurements. Three WLAN sniffers are used in parallel to analyze the packets on channels 1, 6 and 11. The setup is shown in Figure 10. Compared to the interference measurement modifications have been made to monitor all packets and distinguish between packet types (data/management/control). Besides, the monitoring system runs for a longer period of time to obtain a better estimate of the packet type statistics at a particular location. In addition, the measurement setup is installed in a live environment, i.e. no controlled conditions.

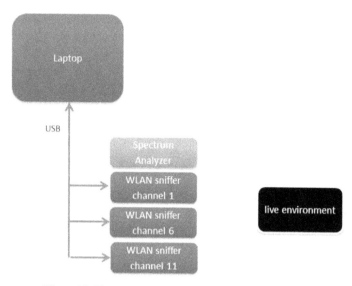

Figure 10 The measurement setup for the live recordings.

5.2 Experiments

Three live situations have been evaluated: college room, office room and city center. The first location, college room, was selected because it is a very crowded place with a large population of WLAN devices in one room. Secondly, in the college room also experiments take place of which some of them use Bluetooth and/or WLAN connections. The second location, an office room, is selected as this is a location which consists of managed WLAN access points, similar the college location. Finally, the third location, city center, was selected as this is a place with unmanaged WLAN access points. Moreover, the place is know for WLAN problems on certain channels.

5.3 Results

Figure 11(a) shows the mean number of frames per second per location. It has been split up into management, control, data and retry frames. First of all the results show that the college room location has most traffic which is due to measurements that were carried out during a college with 75 to 100 students. Secondly in this location about 70% of the traffic are control frames and roughly 21% is actual data traffic. Figure 11(b) depicts the same locations, but here the frames are split up into the most important sub fields.

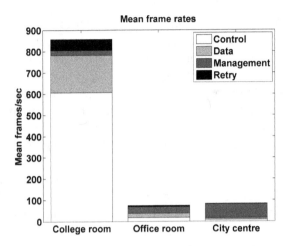

(a) The occurrence of the different type of packets

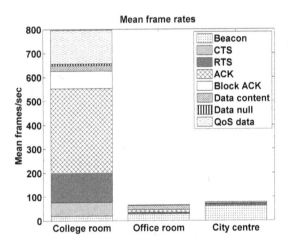

(b) The occurrence of the different subtype of packets

Figure 11 Results for live measurements on three different locations.

Retry frames and less frequent sub field packets are omitted from this figure. This figure reveals that most of the control frames are ACK and Block ACK packets.

Note that the results are in line with the conducted experiments in Section 4.3; here high amounts of Block ACK frames are detected for 802.11e WLAN systems in times of high congestion due to overlapping WLANs.

In addition, Figure 11(b) reveals that only 21% are actual data packets, which is in line with the research conducted by [37]. Moreover, we observe that most of the data frames are QoS frames (16%). This type of frame indicates the use of 802.11e WLAN communication which is more dominantly present compared to the legacy 802.11 WLAN systems (indicated by the regular data content type). Also the figure shows that almost 20% of all traffic are RTS/CTS frames. This means that there is significant interference, probably due to the many WLAN devices. This is also depicted by the retry frame rate (7%) in Figure 11(a). At the office location the mean frame rate is much lower, the actual data frame seems to be similar to the college location, but there are almost no ACK frames. Also the beacon rate is about twice of this rate at the college location. One explanation for this is that in the office location more WLAN networks were active. In the city center, most traffic consists of beacon frames.

College Room Location

The measurements carried out in the college room are most interesting and highlighted below in more detail. These measurements were performed during a college which ended around 1400 seconds. Figure 12 shows the RF spectrum. This is followed by figures showing the occupancy in Figure 13(a), the sub field type versus time in Figure 13(b), and the figure showing the cumulative probability curves for the different types of frames in Figure 13(c). The figures are depicted for channel 1 where during the measurement 119 WLAN devices were identified. The other channels show similar behavior, and for brevity these results are omitted.

From Figure 13(b) it can be seen that the WLAN traffic is very spiky as expected. Interesting enough, when the college ends (at 1400 seconds) a sharp downwards transition is observed in the RF channel occupancy of Figure 13(a). Moreover, at this particular moment the occupancy drops from values around 65 to 20%. The RF monitoring results are supported by the experimental results in Figure 13(b) which shows high traffic in the first monitoring interval (i.e. 0–1400 seconds) and a significant lower traffic load in the second monitoring interval of 1400–2100 seconds. The first interval

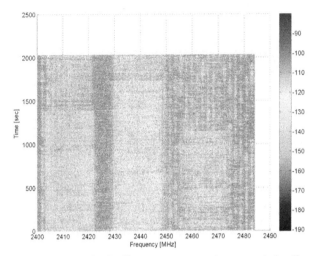

Figure 12 RF monitoring results for live measurements in a crowded college room (field strength values in dBm).

entails high amounts of RTS and Block ACK frames which is not the case for the second interval. Another observation is that the amount of legacy 802.11 data packets is more strongly present during the second monitoring interval in Figure 11(b). However, the amount of ACK frames seems to be steady throughout the entire monitoring session (0–2100 seconds). Interesting, the results in Figure 13(b) roughly indicate that most of the traffic in the highly congested first interval comprises 802.11e traffic which tends to capture channel access. On the other hand, the second interval shows a rise in the legacy 802.11 traffic whereas the portion of 802.11e QoS data and RTS/CTS is considerably low.

To give more insight in the time characteristics the Cumulative Density Function (CDF) curves in Figure 13(c) are provided. Here as a comparison metric, the norm curve represents the situation of constant offered traffic load over time. This figure shows that the QoS data, RTS, CTS and Block ACK CDF curves are constantly positioned above the norm curve. Moreover, the cumulative probability of these packet types is above 80% at the 1400 seconds landmark. On the other hand, the legacy 802.11 data CDF curve displays a flat characteristic during the first interval of high congestion together with a cumulative probability of only 35% at 1400 seconds. However, a steep increase of this type of data packet is observed in the second interval. Thus in line with [35] is seems that 802.11e QoS stations capture the wireless medium in case of congestion, i.e. many WLAN stations. Moreover, this coincides with

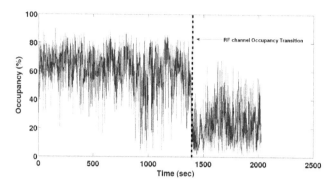

(a) Occupancy of channel 1. At 1400 seconds a significant drop in RF occupancy occurs which is marked by the *transition landmark.*

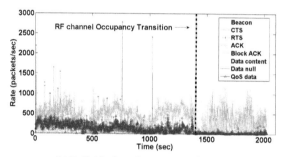

(b) Individual packet rates on channel 1.

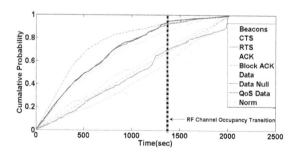

(c) The CDF curves for the different packet sub-types on channel 1.

Figure 13 Results for live measurements in a crowded college room on channel 1. The college ends at 1400 seconds. In the first interval of high congestion ({0–1400} seconds) most of the traffic load is associated with 802.11e WLANs which comprises RTS, QoS data, and Block ACK frames. In the second interval ({1400–2100} seconds) most of the traffic entails the legacy 802.11 type of frames, i.e. data content and ACK frames.

high amounts of RTS/CTS and Block ACK packets which are therefore good indicators of 802.11e induced congestion. This is confirmed by the sniffer readings which show that the QoS data packets are transmitted in blocks bursts.

6 Conclusions

In this paper we have analyzed the spectrum utilization and congestion of 802.11 networks in the 2.4 GHz ISM band. It can be concluded that it is possible to assess the service level. This approach can be applied to other frequency bands where 802.11 technology is deployed such as the 5 GHz band. A cross-layer approach is provided to measure the spectrum utilization and congestion in this band. For this purpose it turns out that packet sniffing allows to identify congestion and that on the other hand spectrum sensing allows to identify utilization. Two situations are investigated using the latter techniques for monitoring.

First, results are presented on the impact of interference between 802.11e WLAN networks sharing the same radio channel in a controlled environment. The results show that the interfering network leads to severe congestion on the wireless medium which in turn occurs in cycles of respectively inactivity and activity. The situation of interferer activity, i.e. congestion, is caused by the RTS/CTS mechanism since the 802.11e WLAN networks often seem to identify each other as hidden nodes. Finally, the conducted experiments in a controlled environments show that the impact of overlapping 802.11e WLANs is relatively irrespective of distance, as long as the WLANs are in the carrier sense range of each other.

Secondly, monitoring sessions in real uncontrolled environments illustrate that only a small portion is actual data traffic. For instance, the live recordings taken in a college room – a location with up to 100 people – indicate that a significant number of packets is classified as control packets (e.g. RTS/CTS), where in turn the number of actual data content packets is quite low (less then 21%). This is in line with the conducted experimental results of the controlled environment, showing that RTS/CTS significantly degrades the performance when two networks are in range of each other; in addition, Block ACKs are identified as a good indicator of congestion in 802.11e WLAN environments. Moreover, the live readings show that 802.11e WLANs typically capture channel access over legacy 802.11 systems, e.g. by using data block burst transmissions. At two other locations, i.e. an office room and a city center site, the traffic mainly consists of management packets and control packets. It can be concluded that the WLAN devices cannot properly handle interference of other IEEE 802.11(e) networks, which leads to very inefficient use of the radio spectrum.Thus there is room for further improvement for more efficient use of the radio spectrum. Further research is needed how this commodity wireless standard can be made more efficient.

6.1 Future Research

It turns out that interference mechanisms are complex and often unexplored in practical situations. Therefore the following topics are interesting for further research. First it is interesting to look at the mechanisms causing interference between WLAN clients of the same network due to packet collisions. Secondly, many open questions remain regarding the interaction between 802.11e and legacy 802.11 systems with respect to for instance overlapping WLANs, e.g. the high amount of Block ACKs and the 20/40 MHz protection mode operation. Moreover, additional research could aim at many WLAN networks on the same channel, and interfering WLAN networks on adjacent overlapping WLAN channels. Thirdly, it is interesting to investigate how the packet rate on the MAC link layer and the spectrum utilization are related and whether an analytical expression can be provided. The relation should take into account the different packet sizes that are possible (e.g. an ACK packet is in general significantly smaller than a data content packet). Finally, to improve on the 802.11(e) standard we recommend adjustments w.r.t. to the RTS/CTS protocol in order to cope with the high amount of RTS/CTS packets once two APs are in range of each other. In line with this, we recommend further inspection of the CSMA/CA protocol so that neighboring APs could share the channel more fairly including improvements on the back-off procedure.

References

[1] http://www.crfs.co.uk.
[2] IEEE Std 802.11e/D13.0, draft supplement to standard for telecommunications and information exchange between systems-LAN/MAN specific requirements. part 11: Wireless Medium Access Control (MAC) and Physical Layer (PHY) specications: Medium access control (MAC) enhancements for quality of service (QoS). IEEE Std P802.11e/D13.0, April 2005.
[3] IEEE standard for information technology-telecommunications and information exchange between systems-local and metropolitan area networks-specific requirements – Part 11: Wireless LAN Medium Access Control (MAC) and Physical Layer (PHY) spe-

cifications. IEEE Std 802.11-2007 (Revision of IEEE Std 802.11-1999), pages C1–1184, 12, 2007.

[4] K. Puttaswamy, H. Lundgren, K. Almeroth, A. Sharma, R. Raghavenda, and E. Belding-R. Experimental characterization of interference in a 802.11g wireless mesh network. Technical Report, University California Santa Barbara, December 2005.

[5] G. Bianchi. Performance analysis of the IEEE 802.11 distributed coordination function. *IEEE Journal on Selected Areas in Communications*, 18(3):535–547, March 2000.

[6] G. Bianchi, I. Tinnirello, and L. Scalia. Understanding 802.11e contention-based prioritization mechanisms and their coexistence with legacy 802.11 stations. *IEEE Network*, 19(4):28–34, July/August 2005.

[7] Sunghyun Choi, J. del Prado, Sai Shankar N, and S. Mangold. Ieee 802.11e contention-based channel access (EDCF) performance evaluation. In *Proceedings of IEEE International Conference on Communications, ICC'03*, Vol. 2, pages 1151–1156, May 2003.

[8] Yoonjae Choi, Byeongjik Lee, Jinsuk Pak, Icksoo Lee, Hoseseung Lee, Jangkyu Yoon, and Kijun Han. An adaptive TXOP allocation in IEEE 802.11e WLANs. In *Proceedings of the 6th WSEAS International Conference on Electronics, Hardware, Wireless and Optical Communications*, pages 187–192. World Scientific and Engineering Academy and Society (WSEAS), Stevens Point, WI, 2007.

[9] Deyun Gao, Jianfei Cai, and King Ngi Ngan. Admission control in IEEE 802.11e wireless LANs. *Network, IEEE*, 19(4):6–13, July/August 2005.

[10] Ilkka Harjula, Jarno Pinola, and Jarmo Prokkola. Performance of IEEE 802.11 based WLAN devices under various jamming signals. In *Proceedings of Military Communications Conference, MILCOM'11*, pages 2129–2135, November 2011.

[11] Guido R. Hiertz, Lothar Stibor, Joerg Habetha, Erik Weiss, and Stefan Mangold. Throughput and delay performance of IEEE 802.11e wireless LAN with block acknowledgments. *Proceedings of 11th European Wireless Conference 2005 – Next Generation Wireless and Mobile Communications and Services (European Wireless)*, pages 1–7, April 2005.

[12] Amit P. Jardosh, Krishna N. Ramachandran, Kevin C. Almeroth, and Elizabeth M. Belding-Royer. Understanding congestion in IEEE 802.11b wireless networks. In *Proceedings of the 2005 Internet Measurement Conference*, pages 279–292, 2005.

[13] Amit P. Jardosh, Krishna N. Ramachandran, Kevin C. Almeroth, and Elizabeth M. Belding-Royer. Understanding link-layer behavior in highly congested IEEE 802.11b wireless networks. In *Proceedings of the 2005 ACM SIGCOMM workshop on Experimental Approaches to Wireless Network Design and Analysis, E-WIND'05*, pages 11–16. ACM, New York, 2005.

[14] S. Khurana, A. Kahol, and A.P. Jayasumana. Effect of hidden terminals on the performance of IEEE 802.11 mac protocol. In *Proceedings of 23rd Annual Conference on Local Computer Networks, LCN'98*, pages 12–20, October 1998.

[15] EunKyung Kim and Young-Joo Suh. ATXOP: An adaptive TXOP based on the data rate to guarantee fairness for IEEE 802.11e wireless LANs. In *Proceedings of IEEE 60th Vehicular Technology Conference*, VTC2004-Fall 2004, Vol. 4, pages 2678–2682, September 2004.

[16] K. Kosek, M. Natkaniec, L. Vollero, and A.R. Pach. An analysis of star topology IEEE 802.11e networks in the presence of hidden nodes. In *Proceedings of International Conference on Information Networking, ICOIN'08*, pages 1–5, January 2008.

[17] Hyewon Lee, I. Tinnirello, Jeonggyun Yu, and Sunghyun Choi. Throughput and delay analysis of IEEE 802.11e block ACK with channel errors. In *Proceedings of 2nd International Conference on Communication Systems Software and Middleware, COMSWARE'07*, pages 1–7, January 2007.

[18] A. Lindgren, A. Almquist, and O. Schelen. Evaluation of quality of service schemes for IEEE 802.11 wireless LANs. In *Proceedings of 26th Annual IEEE Conference on Local Computer Networks, LCN'01*, pages 348–351, 2001.

[19] Anders Lindgren, Andreas Almquist, and Olov Schelen. Quality of service schemes for IEEE 802.11 wireless LANs – An evaluation, 2003.

[20] Jakub Majkowski and Ferran Casadevall Palacio. Dynamic TXOP configuration for QoS enhancement in IEEE 802.11e wireless LAN. In *Proceedings of International Conference on Software in Telecommunications and Computer Networks, SoftCOM'06*, pages 66–70, October 2006.

[21] S. Mangold. IEEE 802.11e: Coexistence of overlapping basic service sets (invited paper). In *Proceedings of the Mobile Venue'02*, pages 131–135, Athens, Greece, May 2002.

[22] S. Mangold, L. Berlemann, and G. Hiertz. QoS support as utility for coexisting wireless LANs. In *Proceedings of the International Workshop on IP Based Cellular Networks, IPCN*, Paris, France, April 2002.

[23] S. Mangold, J. Habetha, S. Choi, and C. Ngo. Coexistence and interworking of IEEE 802.11a and ETSI BRAN HiperLAN/2 in multihop scenarios. In *Proceedings 3rd IEEE Workshop on Wireless Local Area Networks*, Boston, MA, September 2001.

[24] Stefan Mangold, Sunghyun Choi, Peter May, and Guido Hiertz. IEEE 802.11e – Fair resource sharing between overlapping basic service sets. In *Proceedings of the PIMRC 2002*, pages 166–171. IEEE, 2002.

[25] Stefan Mangold, Sunghyun Choi, Peter May, Ole Klein, Guido Hiertz, Lothar Stibor, Cf poll contention, and free poll. IEEE 802.11e wireless LAN for Quality of Service.

[26] Arunesh Mishra, Vivek Shrivastava, Suman Banerjee, and William Arbaugh. Partially overlapped channels not considered harmful. In *Proceedings of the Joint International Conference on Measurement and Modeling of Computer Systems, SIGMET-RICS'06/Performance'06*, pages 63–74. ACM, New York, 2006.

[27] S. Mukherjee, Xiao-Hong Peng, and Qiang Gao. QoS performances of IEEE 802.11 EDCA and DCF: A testbed approach. In *Proceedings of 5th International Conference on Wireless Communications, Networking and Mobile Computing, WiCom'09*, pages 1–5, September 2009.

[28] Qiang Ni. Performance analysis and enhancements for IEEE 802.11e wireless networks. *IEEE Network*, 19(4):21–27, July/August 2005.

[29] Ioannis Papanagiotou, Georgios S. Paschos, and Michael Devetsikiotis. A comparison performance analysis of QoS WLANs: Approaches with enhanced features. *Advances in MM*, 2007.

[30] Nikos Passas, Dimitris Skyrianoglou, and Panagiotis Mouziouras. Prioritized support of different traffic classes in IEEE 802.11e wireless LANs. *Comput. Commun.*, 29:2867–2880, September 2006.

[31] A. Salhotra, R. Narasimhan, and R. Kopikare. Evaluation of contention free bursting in IEEE 802.11e wireless LANs. In *Proceedings of IEEE Wireless Communications and Networking Conference*, Vol. 1, pages 107–112, March 2005.

[32] Burak Simsek and Katinka Wolter. Improving the performance of IEEE 802.11e with an advanced scheduling heuristic. In *EPEW'06*, pages 181–195, 2006.

[33] T. Suzuki, A. Noguchi, and S. Tasaka. Effect of TXOP-bursting and transmission error on application-level and user-level QoS in audio-video transmission with IEEE 802.11e EDCA. In *Proceedings of IEEE 17th International Symposium on Personal, Indoor and Mobile Radio Communications*, pages 1 –7, September 2006.

[34] J. Sydor. Coral: A wifi based cognitive radio development platform. In *Proceedings of 7th International Symposium on Wireless Communication Systems (ISWCS)*, pages 1022–1025, September 2010.

[35] David J. Thuente, Benjamin Newlin, and Mithun Acharya. Jamming vulnerabilities of IEEE 802.11e. In *Proceedings of IEEE Military Communications Conference, MILCOM'07*, pages 1–7, October 2007.

[36] Eduard Garcia Villegas, Elena Lopez-Aguilera, Rafael Vidal, and Josep Paradells. Effect of adjacent-channel interference in IEEE 802.11 WLANs. In *Proceedings of 2nd International Conference on Cognitive Radio Oriented Wireless Networks and Communications, CrownCom'07*, pages 118–125, August 2007.

[37] A.J. Wagstaff. Estimating the utilisation of key licence-exempt spectrum bands. Technical Report Issue 3, Mass Consultants Ltd, April 2009.

Biography

Jan-Willem van Bloem received his M.Sc. in Electrical Engineering from the University of Twente in 2007. He did his internship for British Telecommunications (UK) to work on the BT-FON project in 2006. Since 2010 he follows a PhD trajectory at the Signals and Systems Group of the University of Twente. As part of this trajectory he conducts research for the Radio Communication Agency Netherlands to investigate the service level in the ISM band and the GSM band. Also as part of this trajectory he was employed as researcher at British Telecommunications from October 2011 until April 2012 to work on white-space database assisted collaborative sensing in the TV band.

Roel Schiphorst received his M.Sc. degree with honors in Electrical Engineering from the University of Twente in 2000 for his research on Software-Defined Radio. In 2004 he received the PhD degree for his research on Software-Defined Radio for WLAN standards. Since then, he is a researcher of the chair Signals and Systems. He was also project leader of large T-DAB field trial in Amsterdam which was on-air from 2004 until 2006. His research interests include Software-Defined Radio, digital broadcast systems and digital signal processing for wireless applications.

Taco Kluwer is an innovation advisor for the Radio Communication Agency in the Netherlands. He is currently managing a project where advanced spectrum monitoring and analyses is combined with business intelligence. He received his M.Sc. degree in Electrical Engineering from the University of Twente in 2001, and his research was focused on software defined radio and smart antennas. Since then he has worked as a market specialist and senior technical specialist in numerous projects in the field of radio spectrum monitoring, spectrum management and telecommunication regulation.

Cornelis H. Slump received the M.Sc. degree in Electrical Engineering from Delft University of Technology, Delft, The Netherlands in 1979. In 1984 he obtained his Ph.D. in physics from the University of Groningen, The Netherlands. From 1983 to 1989 he was employed at Philips Medical Systems in Best as head of a predevelopment group on medical image processing. In 1989 he joined the Network Theory group from the University of Twente, Enschede, The Netherlands. His main research interest is in digital signal processing, including realization of algorithms in VLSI.

Author Index, Volume 2 (2012)

Keyword Index, Volume 2 (2012)

Online Manuscript Submission

The link for submission is: www.riverpublishers.com/journal

Authors and reviewers can easily set up an account and log in to submit or review papers.

Submission formats for manuscripts: LaTeX, Word, WordPerfect, RTF, TXT.
Submission formats for figures: EPS, TIFF, GIF, JPEG, PPT and Postscript.

LaTeX

For submission in LaTeX, River Publishers has developed a River stylefile, which can be downloaded from http://riverpublishers.com/river_publishers/authors.php

Guidelines for Manuscripts

Please use the Authors' Guidelines for the preparation of manuscripts, which can be downloaded from http://riverpublishers.com/river_publishers/authors.php

In case of difficulties while submitting or other inquiries, please get in touch with us by clicking CONTACT on the journal's site or sending an e-mail to: info@riverpublishers.com

www.ingramcontent.com/pod-product-compliance
Lightning Source LLC
LaVergne TN
LVHW012331060326
832902LV00011B/1830